TIME, EMBODIMENT AND THE SELF

Beginning with a sustained argument against the new tenseless theory of time and against McTaggart's A series/B series distinction, the author of this essay goes on to provide a non-paradoxical, tensed, phenomenologically-based account of the 'going on' or 'taking place' of events in time that escapes the paradoxes endemic to 'passage' as understood via the A series/B series distinction.

The author then turns his attention to the other main aim of the essay, which is to seek an understanding of time adequate to those more 'embodied' conceptions of the self that place character, and with it the 'constitutive attachments' or 'ground projects' of individual life circumstance, at the centre of the self. This involves a 'redrawing' of the self informed by a wider conception of the will than the one we have inherited via Descartes and Kant, by an account of ground projects, and by the theory of the tripartite psyche in Plato's *Republic*. It also involves extending the account of time developed in the second chapter in a way that draws on the notion of 'ecstatic temporality' that originates with Heidegger.

The essay will be of use to philosophers and advanced students interested in the nature of the self, time, temporality, and phenomenology.

ASHGATE NEW CRITICAL THINKING
IN PHILOSOPHY

The *Ashgate New Critical Thinking in Philosophy* series aims to bring high quality research monograph publishing back into focus for authors, the international library market, and student, academic and research readers. Headed by an international editorial advisory board of acclaimed scholars from across the philosophical spectrum, this new monograph series presents cutting-edge research from established as well as exciting new authors in the field; spans the breadth of philosophy and related disciplinary and interdisciplinary perspectives; and takes contemporary philosophical research into new directions and debate.

Series Editorial Board:

Time, Embodiment and the Self

ANDROS LOIZOU
University of Central Lancashire

Routledge
Taylor & Francis Group

LONDON AND NEW YORK

First published 2000 by Ashgate Publishing

Reissued 2018 by Routledge
2 Park Square, Milton Park, Abingdon, Oxon OX14 4RN
711 Third Avenue, New York, NY 10017, USA

Routledge is an imprint of the Taylor & Francis Group, an informa business

Publisher's Note
The publisher has gone to great lengths to ensure the quality of this reprint but points out that some imperfections in the original copies may be apparent.

Disclaimer
The publisher has made every effort to trace copyright holders and welcomes correspondence from those they have been unable to contact.

A Library of Congress record exists under LC control number: 00131627

ISBN 13: 978-1-138-71293-5 (hbk)
ISBN 13: 978-1-138-71292-8 (pbk)
ISBN 13: 978-1-315-19990-0 (ebk)

Contents

Preface

My central aim in this essay has been to bring together two topics or themes: on the one hand time, or 'temporality', and on the other, the nature of the self. For the most part, the two are pursued separately by philosophers who are broadly within the analytical tradition. In this preface, I shall comment briefly on both. I begin with time.

The discussion of time among analytical philosophers centres mostly around a particular issue, namely that between 'tenseless' and 'tensed' conceptions of time. While a cursory glance will show that I reject the tenseless conception, and that I give some space to providing arguments against it, I want to emphasise at the outset that I am not party to the debate between tenseless and tensed theorists as that debate is currently pursued. As will become clear in the course of chapter one, I reject, for what I hope will be seen as good reasons, the terms themselves in which the debate is pursued, namely McTaggart's A series/B series distinction. In short, the main error that McTaggart's distinction imports - hence an error shared by tensed and tenseless theorists alike - lies in the way that the terms 'past', 'present' and 'future' are thought of as ascribing properties or qualities (of 'pastness', 'presentness' etc.) to antecedently individuated events.

Having produced my arguments against this prevailing discourse in the first chapter, I aim in the second to develop a radical alternative. In the course of developing this new perspective, I refer to Husserl, Heidegger, Sartre and Merleau-Ponty. I have not tried, nor would it serve my purpose to try, to give any detailed exegesis of these philosophers - I have simply drawn on them in support of my overall aims, but without, I hope, too much infidelity to theirs. The crucial chapters are two and four.

The other main topic of this essay, the nature of the self, is pursued mainly in the third chapter. Here I can be identified straightforwardly with one side of an ongoing debate. I endorse a conception of the self to which character, 'constitutive attachments', 'categorical desires' etc. are central, as against the 'thin' or 'unencumbered' conceptions of the self that are a central feature of the epistemological tradition from Descartes through to Kant and beyond. Where I have something new to add here lies in the wider conception of the will (mainly in section 5 of chapter three), and in the discussion of the tripartite division of the psyche in Plato's *Republic* (section 7 of chapter three).

In the final chapter, I bring the two themes together. My overall conclusion is, roughly stated, that the conception of time that is germane to the standpoint of the observer - for example, time as it is presented to us when we are in 'listening' mode, or when we are spectators of a theatrical or a musical performance - is grounded in the more basic temporality of our prereflective engagement in the world, the world of our 'constitutive attachments'. My hope is that the notion of the intra-temporal 'I' to which I speak in section 4 of the final chapter will throw some light on the discussion of the self in the third chapter, and perhaps serve to add weight to the move to reinstate character at the centre of the self.

Finally, I would like to acknowledge the sources of help I have received, directly and indirectly. Chief among these are my students over the last five years, especially those who in their third year were willing to 'brave the rapids' with me. I am also grateful to Jonathan Lowe for reading drafts at various stages of completion, for some electronic correspondence, and for his general encouragement over the last few months. I would also like to thank my colleague Doris Schröder for helping me with some of the technicalities of production.

Andros Loizou

1 McTaggart's Parody of Time, and the Tenseless Theory

1 Introduction

The difference between the discussion of time in such texts as Heidegger's *Being and Time* or Merleau-Ponty's *Phenomenology of Perception* and in the works of philosophers belonging in the British analytical tradition is, on any account of the matter, striking. Through the one tradition, the continental European, an understanding of time (or 'temporality') is developed which seeks to illuminate, or at least address, some central issues of human existence, the nature of the self etc. in ways which seem at best fanciful, at worst incomprehensible, to philosophers in the analytical tradition. Discussion of time in the analytical tradition, by contrast, has most recently been highly technical, narrowly focused, and for the most part centred round a particular controversy which seems *very far removed* from any such central issues. The student of philosophy who wishes to understand this controversy among analytical philosophers has, as his starting-point, ordinary everyday speech; but if he wants to make any progress, he must be ready to make some departures from everyday speech and accept a particular hybrid construction of it.

This hybrid language originated with the British idealist philosopher J.M.E. McTaggart.[1] The irony is that whilst for McTaggart this hybrid language served only to 'prove' the unreality of time, for the analytical tradition following the demise of British idealism it has become the unquestioned backdrop against which the central debate over the nature of time is pursued - so much so that almost all parties to the debate no longer see it as hybrid. But this has not always been so in the analytical tradition. There have been philosophers who have questioned McTaggart's language of time, such as C.D. Broad and G.E. Moore in an earlier period, and David Pears in the 1950s.[2] It is my claim that this hybrid language is deeply flawed, and that the ways in which it is flawed are instructive. To bring out these flaws will be the initial aim of this first chapter. But the main aim will be to provide some compelling arguments against the major variants of the popular and widely-supported theory of time known as 'the new tenseless theory'.

Starting with chapter 2 of this essay, I shall begin to introduce into the discussion some of the key texts in the phenomenological tradition which have a direct bearing on time and temporality. That tradition, as I hope to show, has much to offer us.   But I also hope to show how this understanding can be further enhanced, both by attending to the flaws in McTaggart's hybrid language of time, and by taking note of some of the insights and methods of analytical philosophy generally.

I begin, then, with McTaggart.

2 McTaggart's Language of Time and its Inconsistencies

The hybrid language bequeathed by McTaggart is, of course, the now famous distinction between what he calls 'the A series' and what he calls 'the B series'. The bequest was not altogether a happy one, as I have argued elsewhere.[3] I shall not repeat that earlier discussion - but because McTaggart's language has exercised such an influence on English language analytical philosophy, and continues to do so, something needs to be said by way of summary.

McTaggart introduces the A series/B series distinction in the following way:

> Positions in time, as time appears to us *prima facie*, are distinguished in two ways. Each position is Earlier than some and Later than some of the other positions. To constitute such a series there is required a transitive asymmetric relation, and a collection of terms such that, of any two of them, either the first is in this relation to the second, or the second is in this relation to the first. We may take here either the relation of 'earlier than' or the relation of 'later than', both of which, of course, are transitive and asymmetrical. If we take the first, then the terms have to be such that, of any two of them, either the first is earlier than the second, or the second is earlier than the first.
>
> In the second place, each position is either Past, Present or Future. The distinctions of the former class are permanent, while those of the latter are not. If M is ever earlier than N, it is always earlier. But an event, which is now present, was future, and will be past.[4]

The tenseless (earlier/later) series McTaggart called 'the B series', and the tensed (past present and future) series he called 'the A series'. Now to anyone looking on these linguistic devices from outside the debate among analytical philosophers, a number of questions must arise: Are these two distinct *series*? Or is there just the *one* time series, presenting, so to speak, two different *aspects* of time? The reader soon discovers that McTaggart intends the latter. But the question why we should need *both* series is likely to remain.

To the intelligent layman, the A series may appear to be all that is needed. His initial reading of the two ways in which 'positions in time' are supposed to present themselves is likely to be this: the B series secures for us the *order* of events - for as McTaggart says, B series positions do not change, and 'if M is ever earlier than N, it is always earlier' - whereas the A series captures the fact that events 'flow' or 'pass' from the future, into the present and thence into the past. So, he or she might now think, the B series embodies temporal *order*, whereas the A series embodies temporal *flux*.

Despite its obvious attractions, this does not stand up. The A series is, after all, a *series*, and not merely a set of qualitative distinctions - it is *'that series of positions* which runs from the far past through the near past to the present, and then from the present through the near future to the far future, or conversely'.[5] The A series does not, therefore, represent temporal flux *alone* - whatever this could possibly mean - but includes within it, within the overall flux or movement, a serial ordering of events (or 'positions') as well. So our intelligent layman might now legitimately ask: Given that the A series already contains a serial ordering of events or 'positions', *why do we need the B series at all?*

We distinguish succession from duration, the flux from the abiding or permanent, or (as Broad puts it) between the 'transitory' and the 'extensive' aspects of time.[6] But none of this is of any avail if we want to preserve McTaggart's distinction. Only by modifying McTaggart's distinction in the way Gale does - i.e. by transforming McTaggart's A series into what he (Gale) calls 'the pure A series' - is it possible to align McTaggart's A series/B series distinction with such distinctions as transitory/extensive, fleeting/abiding, succession/duration etc.. 'The pure A series' consists simply in distinguishing past, present and future from each other in a purely *topological* way, i.e. with no *series* of degrees of pastness and futurity.[7] One can then identify Gale's 'pure A series' with the transitory aspect of time, leaving the extensive aspect to the B series. The A series is now reduced to something well captured in the metaphor of a spotlight moving along a fixed track - the spotlight of 'presentness' passing to later and later terms of the firm, fixed track of the B series. The separation of the two aspects of time is then complete, but at a price - perhaps too high a price, as we no longer have *McTaggart's* A series. But as I argued in the earlier essay, for such a misconstrual of McTaggart's A series to be *on the cards at all* suggests that it is inherently prone to this misconstrual - suggesting perhaps that, just as Gale's 'pure A series' is an attempt to separate the transitory from the extensive aspect of time, so too McTaggart's distinction itself constitutes a somewhat confused step in that direction.[8]

There is something peculiarly paradoxical about some of the ways in which McTaggart represents the relation between the A series and the B series. In a footnote, he says that we either think of the A series as sliding along the B series, or conversely we think of the B series as sliding along a fixed A series: 'The movement of the A series along the B series is from earlier to later. The movement of the B series along the A series is from future to past'.[9] This suggests the image of a track with its fixed points, and a measuring-rod of indefinite length sliding along it; or, alternatively, a reel of film, illuminated and projected onto the cinema screen one frame at a time. Whichever alternative we take, the upshot is the same - the changes of tense determination which alone can constitute the passage of time are no longer represented as internal to the A series, but as changing relations between the A series and the B series. On both, we have not only 'now', but other fixed tense determinations, such as 'yesterday', 'tomorrow', 'two-days-ago', 'last month', 'a year hence' etc. passing to later and later terms of the B series. But in speaking this way, McTaggart has repeated himself. The A series as defined initially is *not* a series of fixed determinations of this kind, but of *changing* determinations - for it is already *internal* to it that the events of tomorrow should change a day hence and so come to bear the determination 'today' or 'now'. We do not therefore need to set up a 'motion' between the A series and something outside it, since it *already contains* its own 'motion' - thus making the B series redundant.

In a footnote to the *Mind* 1908 article, McTaggart speaks differently of the 'movement' of events: 'If the events are taken as moving by a fixed point of presentness, the movement is from future to past....If presentness is taken as a moving point successively related to each of a series of events, the movement is from past to future'.[10] Here McTaggart comes close to Gale's 'pure A series', and hence to the separation of the 'transitory' and 'extensive' aspects of time I discussed earlier.

McTaggart is far from clear on how he conceives the relation between his A and B series, and on whether he considers the A or the B series as more fundamental. After introducing the A series/B series distinction, McTaggart says: 'It is clear ...that, in present experience, we never observe events in time except as forming both these series'.[11] This remark, taken by itself, might seem to suggest that the two series are equally important. But then he says that the A series is *more fundamental than* the B series, in the sense that if there were no A series - if the A series were shown to be unreal - there would be no B series. Yet he does not seem to think (he certainly does not say) that *if there were no B series*, there might still nonetheless be an A series. Indeed, all the indications point to his thinking it an obvious truth that anything that was in an A series would *ipso facto* be in a B series.[12] In what way, then, is the A series but not the B series 'fundamental'? And then, when McTaggart introduces the C series - which, for present purposes, we can think of as an uninterpreted ordered series, or 'bare seriality' as such[13] - he says that although the A series and the B series are equally *essential* to time, it is only the A series and the C series that are *ultimate* - we are told that the B series is *derivative*. In the face of these different assertions and the seemingly unresolvable tensions between them, it is hard to see what McTaggart's B series really amounts to; and so it is hard to decide whether an event's changing tense-determinations are to be understood through the relations in which the A series stands to the B series, or through the A series alone. The ordinary man innocent of philosophy might be forgiven for concluding that the B series is a fiction, given that the A series as originally defined by McTaggart - i.e. as an *extended* series of *changing* tense determinations - can accommodate the extensive as well as the transitory aspect of time.

However, given McTaggart's own account of the A series and the way he first introduces it, even if we tried to banish the B series we would be compelled to reinvent it. For McTaggart's account of the A series is, precisely, one which invites the image of events on a moving conveyor belt, or that of the boat sailing down the canal past the row of houses on the bank. Furthermore, it is an account in which the future is not characterised as 'open' in an ontological sense, but as determinately in being, 'closed'. The death of Queen Anne, he says, was an event with its determinate character 'before the stars saw one another plain'. But in the following respect it does change: 'It was once an event in the far future. It became every moment an event in the nearer future. At last it was present'.[14] McTaggart sometimes speaks of these 'changes' as if they are fortuitous: 'Changes must happen to the events of such a nature that the occurrence of these changes does not hinder the events from being events, and the same events, both before and after the change'.[15] The temptation will always be there, in other words, to reintroduce the 'spotlight' image of an A series - Gale's 'pure' A series - and to view the eternal order of events in terms of the (perforce reintroduced) B series. The 'pure' A series would then come to seem ever more and more ephemeral, and the B series ever more and more solid. In summary, the fact that the A series is ontologically closed in the same way as the B series, together with the fact that McTaggart speaks of changes of tense determination *happening to* the events in such a way as to leave them fundamentally unchanged, suggests that even if we denied the external existence of the B series, we would have to reinvent it from within the A series itself.

3 Tenseless Theory and the Future

It is easy to see why, given McTaggart's way of characterising the passage of time, some philosophers lose patience with the idea altogether and are drawn into embracing a tenseless theory. McTaggart's 'account' of the passage of time is a parody of it. This should not be surprising, given that his purpose is to show time to be unreal. Those not attracted to the tenseless theory must face the task of rewriting the metaphysics of time, and of the passage of time, in a way which is no longer hostage to the language bequeathed to us by McTaggart, and to the paradoxes endemic to it. This requires, among other things, giving up the idea of 'past', 'present' and 'future' as *properties or qualities of* something, e.g. events. I return to this later, after looking at some problems with the tenseless theory.

The merit of the tenseless theory is its simplicity, and simple theories do have a certain elegance. Repelled by the paradoxes which the idea of past, present and future as distinctions qualifying reality itself generates, the tenseless theory seems a safe haven - all the more so, perhaps, when (as is claimed for it in its new form) it provides us with an explanation of why we need to invoke these distinctions. The distinctions of past, present and future are then relegated to the context of belief ('we cannot help having tensed beliefs', as Mellor puts it) leaving reality itself essentially tenseless. The theory is simple: a particular token of a tensed sentence - e.g. 'Jim will race tomorrow' - is true just in case it occurs earlier than the time t designated by a true tenseless statement of the form 'Jim races(tenselessly) at t'. And the latter statement is *omnitemporally* true, i.e. not just true at t.

The central tenet of the tenseless theory of time is that 'tensed statements have tenseless truth-conditions' - an innocent-sounding claim on the face of it, but one which has some very counter-intuitive consequences. Among these is the tenseless theorist's commitment to the tenseless, hence omnitemporal, truth of statements about future events. But there are strong grounds for thinking that this particular commitment carries with it the failure to acknowledge that we need two notions of possibility: an ontological as well as an epistemological one. The tenseless theorist may well be committed to accepting only the one, namely the epistemological notion. A recent exponent of the (new) tenseless theory certainly takes this view:

> If tenseless theory is correct, then one kind of indeterminacy is certainly ruled out: the future cannot be *ontologically indeterminate* or 'open', that is, it cannot just consist of a collection of possible worlds. Future fact is just as much part of reality as past and present fact. But the future is certainly *epistemologically indeterminate....*[16]

In what follows, I hope to bring out two points in connection with this claim: that tenseless theory *does* indeed rule out an 'ontologically indeterminate' or 'open' future, and that ruling this out does entail some unacceptable consequences.

The law of excluded middle is, for tenseless theory, this: (*a*) Either it is (tenselessly or omnitemporally) true that X happens at t, or it is not (tenselessly or omnitemporally) true that X happens at t. In terms of 'statements about future contingents', this translates into (*b*) either it is now inevitable that X will happen tomorrow, or it is now inevitable that X will not happen tomorrow. (It is important to notice where the negation sign in the second disjunct of *b* is placed, and that an alternative placing is possible. I return to this presently.)

Statement *b* is commonly thought to be, or to entail, fatalism. The fatalist goes on to infer, from *b*, the following: If it is now inevitable (or is tenselessly true) that X *will* happen tomorrow, then any attempt to prevent it will be ineffective, and any attempt to bring it about will be superfluous. If, on the other hand, it is now inevitable (or is tenselessly true) that X *will not* happen tomorrow, then any attempt to bring it about will be ineffective, and any attempt to prevent it will be superfluous. The future is thus 'ontologically' closed, 'all laid up', and the only sense in which it is 'open' is that, by and large, we know very little about it, and so we say 'it is possible that X will happen tomorrow', meaning by this merely that it is possible *for all we know* that X will happen tomorrow.

Let us now, for a moment, step outside tenseless theory and consider a conversation between a fatalist and myself (who, against the fatalist, wishes to uphold some notion of an 'ontologically' open future). The fatalist begins, 'either it is inevitable that X *will* happen tomorrow, or it is inevitable that X *will not* happen tomorrow. If X *is* going to happen, then any action to prevent it will be ineffective, and any action to make it happen will be superfluous, redundant. If, on the other hand, X is *not* going to happen, then any action to prevent it happening will be superfluous and redundant, and any action to make it happen will be ineffective. In short, it is *now true* either that X will happen tomorrow or that it will not'. To this I reply, 'when you say that *it is now true* that either X will happen or that it will not, what exactly is the force of your claim that either the future tensed statement, asserted now, is true, or its negation, asserted now, is true? Are you claiming *anything more than* that at some time tomorrow, one or the other of the two present tensed statements "X is happening now" or "X is not happening now" will *then* be true or correctly assertable?'

Much depends on how the fatalist replies to my last question. If he says, simply, 'Yes, that is all I'm saying', then clearly no fatalist consequences will follow at all; if 'is now true/false' as *he* uses this expression, i.e. to qualify the present utterance of a future tense statement, means no more than what I mean when I use the expression 'will be true/ false', i.e. to qualify the future assertion of a present tense statement, then we have the harmless tautology 'whatever will be will be', and not the fatalist thesis 'whatever will be *must now* inevitably be'.

But the chances are, if he is a serious fatalist, that he will not reply thus. He will want to say that his present assertion of a future tense statement means *more* than my future assertion of the corresponding present tense statement, and he will want to say that his position requires no more than the correct application of the law of excluded middle. Let us assume this is his view. I can now challenge him to clarify his reading of the law of excluded middle. What, I ask him, is the negation of 'it is now true or now inevitable that X will happen tomorrow'? He will reply, 'it is now true, now inevitable, that X will *not* happen tomorrow'. But I now offer him an alternative negation of the original statement, viz. 'it is *not* now true, *not* now inevitable, that X will happen tomorrow', and I point out to him that from its not being now inevitable that X will happen tomorrow it does not follow X will not happen, and neither does it follow that it will - and that *my* statement is more naturally construed as the negation of the original statement than *his*. I may then go on further to say that, whilst the present utterance of the future tense proposition 'it is now true or now inevitable that X will happen', if true, logically *entails* the truth of a present tense proposition 'X is happening now' uttered tomorrow, the truth or correct assertability of 'X is happening now' uttered tomorrow does *not* entail the truth of 'it is now true (now inevitable) that X will happen tomorrow' uttered *now*. My reasons for thinking the entailment holds only one way will be clear from what immediately follows.

If the fatalist asks me how *I* would be prepared to construe his future tense proposition, I reply that I can only construe it as making a particular causal claim, namely: If it is *now inevitable* that X will happen tomorrow, this must be because there are now in existence the causally sufficient conditions for making it inevitable that X will happen tomorrow. If there are no causally sufficient conditions in existence, then no sense can attach to the notion of present - or omnitemporal - truths about future contingents *per se. Whether or not* there is such a set of causally sufficient conditions is, in each particular case, an empirical matter, and not one that can be settled *a priori*. Whenever there are *no* such present sufficient conditions binding the future, the future is, in the respects in question, open in an ontological and not a merely epistemological sense. For the fatalist, by contrast, the closedness of the future is an *a priori* fact - as it is for the tenseless theorist, despite any disclaimer that some tenseless theorists may be inclined to make about this.

Tenseless theory thus seems to be flawed on the same grounds as fatalism, and we need to determine how far and in what way this is so. There are two aspects of the tenseless theorist's programme which have a bearing on the issue of the future and which seem irreconcilable. First, he wants to say that tensed statements are made true in every case by tenseless facts. These tenseless facts are typically articulated by means of tenseless statements of the form '*a* is F at *t*', where *a* is a subject term, F a predicate and *t* a date. Such statements are omnitemporally true, hence 'true now'. Second, he wants to say that the foregoing position can accommodate all that we would want to say, in tensed speech, under the heading 'the open future'.

But if the earlier arguments against the fatalist are correct, the tenseless theorist cannot accommodate these things. Future possibilities as *he* understands them are possibilities only with regard to *knowledge*; the future is open only in an epistemological and not an ontological sense. He cannot accept the fact that the negation of 'it is now true (now inevitable) that X will happen tomorrow' is 'it is not now true (not now inevitable) that X will happen tomorrow' - hence, when the latter (the negation) is true, the future is at the time of utterance open in an *ontological* and not a merely epistemological sense, i.e. open in a way that allows for the large commonsense fact that it is *an empirical matter* whether or not it is 'now true' or 'now inevitable' that X will happen tomorrow, a matter of whether or not the causally sufficient conditions for X's happening tomorrow are now in place, and not something to be settled *a priori*.

In summary, then, the claim that the negation of 'it is now true, now inevitable, that X will happen tomorrow' is 'it is not now true, not now inevitable, that X will happen tomorrow' threatens the tenseless theory at its very core. For when the latter, the negation, is true, the future is genuinely open, open in the sense in which there are *no* causally sufficient conditions in place at the time to determine whether or not X will happen tomorrow. It is important, incidentally, to notice that to speak in this way is *not* to abandon the principle of bivalence. We *do not* have to invoke 'three valued logic' or any other such device to account for it. If it is *not* now inevitable that X will happen tomorrow, the law of bivalence operates in a way that 'either X will happen tomorrow or it will not' *means no more than* that when tomorrow comes, either the present tense statement 'X is happening now' or its negation 'X is not happening now' will *then* be true or correctly assertable.

But the tenseless theorist will want to resist this with all his might, because accepting it would mean that there could be true tensed statements about the future *for which no tenseless equivalent can be found* - and this would entail the falsity of his central thesis, viz. the thesis that *all* tensed statements are made true by tenseless facts. It would entail that at least *some* of our tensed beliefs do not have tenseless truth conditions. The tenseless theorist understands the negation of 'it is now true (now inevitable) that X will happen tomorrow' to be 'it is now true (now inevitable) that X will not happen tomorrow'. But there is no good reason to grant him this, and every reason to reject it. When we negate a proposition, we negate the whole proposition, and not an embedded clause. Negating the proposition gives us 'it is not now true (or now inevitable, or omnitemporally true) that X will happen tomorrow'. And yet one or other of the disjuncts 'X is happening now' or 'X is not happening now' will be true, will be correctly assertable, tomorrow. But it is open to me to make a prediction, or a guess, and assert *now* that X will happen tomorrow, or assert *now* that X will not happen tomorrow, but *without* committing myself to the claim that it is 'now inevitable' (or 'now true', or 'omnitemporally true') that X will happen tomorrow (or will not happen, as the case may be). I may even *believe* one or other of these propositions - in which case I will have a belief to which no tenseless fact corresponds, and which therefore does not have tenseless truth conditions. The new tenseless theory therefore fails, and does so in its central thesis.

In case it is felt that this way with the tenseless theorist is too short, in the last part of this section I shall further clarify the basis for this conclusion and at the same time try to meet some possible objections.

Once the tenseless theorist admits that the future can be open only in an epistemological sense, the foregoing conclusion is inevitable. But not all tenseless theorists are ready to admit this; and it is anyway doubtful whether the arguments of the last section will convince the truly committed tenseless theorist. For despite the fact that we *do* use the future tense in the way I have just shown, he may just dismiss this as simply a quirk of usage, or just insist that even here it is possible to speak of tenseless truth conditions. The tenseless theorist would presumably claim that there is a 'truth value link' such that, in the case where there are at present no causal conditions in place to ensure that X will occur tomorrow, and yet in the event X *does* occur tomorrow, the correct assertability tomorrow of 'X is now occurring' *entails* the correct assertability, the truth, of 'X will occur tomorrow' uttered *today*. But the tenseless theorist cannot without circularity invoke this claim as a premise in an argument designed to *establish* the thesis that all tensed statements have tenseless and omnitemporal trurh conditions, since the claim itself only makes sense on the *presupposition* of the thesis.

Hence in relation to my thesis about 'future contingents', the tenseless theory can only appear in a reductionist light. But many tenseless theorists want to deny any reductionist tendency, and resolutely insist that *anything and everything we may want to say* using tensed statements (and here I mean *tokens*, not *types*) is made true by tenseless facts. Very well. If any tenseless theorist wants to insist on this, then he must extend his generosity so as to accommodate my view on future contingents - a view no different from that of the ordinary man or woman innocent of the tenseless theorist's sophistications. If he is *serious* in wanting to accommodate this view on future contingents, then he must answer 'no' to a question I asked my fatalist interlocutor in the last section. That question was whether, in his use of 'it is true (or it is omnitemporally true, or it is now true) that X will happen tomorrow', he is claiming anything more than the future correct assertability of a present tense statement (again a *token*, not a type) 'X is happening now'. But if he *does* answer 'no' to this question, then his notion of the omnitemporal truth of 'B-statements' has to sit, somewhat uncomfortably, beside a notion of the future as *ontologically* open.[17] It is doubtful whether one *can* affirm both.

The upshot is, then, that either the class of omnitemporally true B-sentences is not universally inclusive (i.e. it does not cover all possible cases, since it falls down on future contingents) - which means that tenseless theory is just simply wrong - or, if the tenseless theorist takes a generous stance on future contingents, his theory courts the accusation that it is vacuous, with the notions of B-facts, omnitemporally true B-statements and the rest doing no work at all - like Wittgenstein's example of the knob that looks as if it could turn on part of a machine, but turns out to be a mere ornament not connected to the mechanism at all.[18]

The question I posed to my fatalist interlocutor turns out, in short, to be not an easy one for the tenseless theorist. Whichever way he tries to answer it, there are problems for him. He would rather, I suspect, just repress the question.

If the arguments given here are correct, then tenseless theory simply fails, *whatever* version we consider. It cannot accommodate many of the tensed beliefs we have about the future. Consequently, if embracing tenseless theory is the overriding consideration, we have to abandon those ways of understanding the world which depend, in one way or another, on the supposition that the future can be genuinely open.

In sections 4 and 5, I explore some of these ways of understanding the world that the main body of tenseless theorists hope to preserve, notably the idea of a thing and the distinction between things and events. What I hope to show is that tenseless theory cannot preserve these, and hence if we want to embrace tenseless theory, we have to pay a very high price.

4 Tenseless Theory and Persisting Things

In this section, I take up some points arising out of one of the most recent expositions of the tenseless theory: *Real Time II*, by D.H. Mellor. I shall be contrasting this with the tenseless theory put forward by R. Le Poidevin in *Change, Cause and Contradiction*, which I shall look at in the next section.[19] One important difference between the two is that the latter espouses the thesis that persons and things have temporal parts, whereas the former explicitly denies this.[20] For Mellor, in other words, a table, a chair or a person exists *as a whole* at any time *t*, and not as a succession of temporal parts during successive and contiguous phases of its history. But what I hope to show is that this more catholic view of the matter does not, in fact, square with tenseless theory, which is more intelligible on the thesis that persons and things have temporal parts. (In fact, even the theory of temporal parts is problematic for tenseless theory. I shall return to this in the next section). Tenseless theory, I shall argue, cannot sustain the more catholic notion of thing or person identity because it cannot sustain a number of requirements that go with it: most notably, a *clear* distinction-in-principle between essential and non-essential properties of a thing, the idea of the future as *ontologically* and not merely *epistemologically* open, and - perhaps most important of all - the fact that for an indefinitely wide class of beings or things, including persons, there is what we might call 'context transcendence', or more specifically, 'transportability across space without loss, or threat of loss, of identity'. What this means should be obvious - to spell it out by way of an example, it means that while it happens that at this moment (11.02 am on Sunday 24th April 1999) I am in my study typing this text, I *could* have been instead mowing the lawn in my back garden; and this does not imply that, had this been the case, I would have been a (slightly, even *ever*-so-slightly) different *me* - my identity is strictly non-negotiable, as is that of my desk or my pen.

A glance at some of the central pillars of Mellor's defense of the tenseless theory will make it clear that the theory cannot accommodate the notion of identity such examples require. In particular, his exclusive reliance on causation for providing an account not only of changes in things, but of the very persistence of the things themselves, places too great a burden on that overworked category, a burden it cannot support. Mellor's discussion of persistence and identity is at a number of points like Hume's. But where Hume's strategy for avoiding talk of 'substance' involves speaking of my perception of a table as made up of a succession of similar *impressions*, thus inviting the reply that he is treating a *thing* as if it were a *series of events* (albeit events that are qualitatively similar), Mellor seeks to avoid this by appeal to 'factual causation'. Briefly, a thing A has the same set of properties S at t_2 because it had them at the earlier time t_1, and it has them at t_1 because it had them at the earlier time t - the facts obtaining at t cause the facts obtaining at later t_1, which in turn cause the facts obtaining at the even later time t_2.

It would be wrong to deny that what distinguishes things from events most of the time is the stability, the permanence, of the former. As Mellor notes, many (but not all) events are changes in things. In such cases the events in question owe their identity to the things in which they constitute changes - if my pencil snaps, the event is the snapping of just *this* pencil and not any other.[21] The pencil constitutes the stable, permanent and underlyingly changeless sustainer of the change in question. But is this changelessness of things the *only* important feature of them? Where does the notion I invoked above, viz. 'transportability across space', fit into the picture?

We need to distinguish at least two *kinds* of things in this connection. First, there are those which are not transportable, because in one way or another they owe their identity to their location. These comprise both natural and conventional cases. The Grand Canyon is an example of the former, and Church Road Methodist Church in St. Annes is an example of the latter. Taking the first, there are difficulties in the very *idea* of moving the Grand Canyon to, say, the Gobi Desert - difficulties that are wholly absent from the idea of, say, dismantling a building brick by brick and re-assembling it exactly as it was but in another location. Hence, to turn to the conventional cases, there would be nothing wrong in principle with dismantling *the actual building* which constitutes the church and reassembling it in California - but this would be simply the re-location of a building, not of Church Road Methodist Church as such, because the latter owes its identity to the Methodist community of which it is the church, and they could simply set about erecting a different building for their purposes.

The second kind of things is best exemplified by the vast range of human artifacts - tools, building materials, items of equipment generally, furniture, books, cars, bicycles, sculptures and paintings, but it also includes whatever we appropriate from nature, namely timber, water, rocks, soil, sand, animals and so on. Now things of this kind have two important characteristics: they are the *substrata* of change, and so they are in essential respects *un*changing; and they are transportable without loss of identity as mentioned previously. And we normally think of these characteristics as inseparable - yet they are clearly distinct, or we would not be able to speak at all of the kind of things discussed in the last paragraph, namely essentially *un*transportable things such as the Grand Canyon, the Ribble Valley, Mount Everest and certain local institutions (a community, a region and so on). The Ribble Valley, for example, is in the required sense unchanging, yet it is not transportable. That things of *this* kind are unchangeable substrata or sustainers of changes constitutes both a necessary and a sufficient characterisation of them. But could the converse hold? That is, given that the first kind of things is characterised by unchangeability but lacking transportability, could we conceive of anything belonging to the second kind of things that was transportable in the required sense, but not unchanging?

I shall return to this presently. But first, let us consider a prior question, namely: could we conceive of a world comprising things *only* of the first kind, i.e. to the total exclusion of things of the second kind? This is, I confess, a rhetorical question, for I certainly could not make sense of such a world - a world in which the slightest movement radically altered the identity of everything. I could not get out of the house knowing that my paperback thriller would still be the same, let alone rely on it to make my train journey more bearable. (How could 'a book' *exist at all* in such a world?) Clearly things of the first kind are made up of, or otherwise contain, things of the second kind. I can bring back a piece of Mount Everest, or of the Grand Canyon, as a souvenir - a clod of soil in the first case, or a piece of coloured rock in the second. So transportability, and hence the existence of things of the second kind, is a necessary feature of the world. We could not conceive of a material spatio-temporal world, in other words, in which things of the first kind were not made up of things of the second kind - Mount Everest of transportable bits of snow, or of rock, or of soil, and the Grand Canyon of transportable pieces of rock of different colour etc..

Let us now consider whether it is possible to conceive of transportable yet *radically changeable* things. What follows is a thought experiment in which I hope to show that they are conceivable, and hence that transportability can precede changelessness as the way things are individuated.

My friend in Yorkshire is an inventor of highly eccentric things. He has invented 'the ipongo' (plural 'ipongoes', as in 'tomatoes'), and he sends one to each one of us, his closest friends. Mine duly arrives, and on opening the parcel I find a green three inch cube on a black plinth, the whole thing looking as if it is made of polished marble. The instructions say, simply, 'place the ipongo anywhere, move it about from place to place as you please, and monitor its behaviour'. After a day or so, I notice it getting warmer to the touch and beginning to glow. Then a few hours later, I find that the cube has been replaced by, or has changed into, a glowing yellow sphere on the familiar black plinth! I move it to another location again. Two days later, I come down to breakfast to find a black horse of roughly the size of the original object, and no plinth. I then embark on an extensive phase of monitoring, in which I establish the following facts: first, the ipongo repeats these three phases at regular intervals, viz. first the green cube on the black plinth, then the yellow sphere on the black plinth, and then the black horse without the plinth; and second, it is transportable *during* the process of change - I witness its change from one phase to the next while I am transporting it from one place to another. (It is part of this second fact, of course, that I duly witness the transition from place to place to be by means of a *continuous* path through space.)

My friend next informs me of 'the ipongo mark two', which is similar to the mark one except that it goes through an endlessly repeated cycle of ten rather than merely three phases. I monitor it and find, as before, that it changes phase at predictably regular intervals, and that I can be moving it during the transitions. But what *really* astonishes me is the advent of 'the ipongo mark three', which, as I discover, has an indefinitely large number of phases, and further, unlike mark one and two, changes unpredictably - although in common with mark one and two, it is transportable during its transitions.

The chief conclusion to be drawn from this is that ipongoes of all three types can be individuated prior to our knowledge of their properties. If the thought experiment succeeds, then it clearly shows that the main criterion governing thing-identity is transportability rather than sameness of properties, unchangeability etc. (Curiously perhaps, the ipongo mark three makes this claim *more* rather than less plausible - for in this case, transportability is all there is to fall back on.)

This conclusion is hardly new. It is at least implicit in Strawson's reconstructed Kantian project in *The Bounds of Sense*. I have in mind here his claim, made throughout the book, that it is not time *alone* that allows us to distinguish the order of our perceptions of the world - the 'subjective' order - from the 'objective' sequence, or lack of sequence, which belongs to the world independently of our perception. A space of three dimensions is also needed, through which it is possible for subjects of experience to take any number of different routes, in turn giving rise to any number of different subjective orderings of their experiences, without prejudice to the objective order.[22] And there is no very great logical distance between this claim, which concerns transportable *subjects* of experience, and the transportability criterion for the identity of *things* adumbrated here.

Clearly these formal requirements can be fulfilled only in a world in which the future is open in an ontological and not merely an epistemological sense - and by this I mean simply a world in which present causal conditions do not close the future totally. When I say that my piano *could* be moved into the lounge rather than the study tomorrow, this is a genuine *de re* or metaphysical possibility, not merely an epistemological one. I am not saying that *for all I know* the piano's future history lies in the lounge rather than the study - implying that I simply lack knowledge of a determinate causal outcome, or a body of determinate 'tenseless fact'. What I am saying is that, at this point in time, there are *no* causes in existence which determine where my piano will be tomorrow, and that it is in an important sense *up to me* where it will be. And this is without prejudice to the fact that tomorrow, one or other of the two propositions 'my piano is in the lounge' or 'my piano is not in the lounge' *will be* true or correctly assertable. It should be noticed that the converse also holds: an ontologically open future that I can rely on in this way clearly requires that there should be enduring or persisting things like pianos which, in the same way as ipongoes, preserve their identity through transportation. These requirements are clearly impossible to square with tenseless theory, which - if it is to amount to a distinguishable view - must presuppose a future that is ontologically closed.

If for these reasons Mellor's avowal of an ontology of persisting things is difficult to make sense of, the account he actually gives of them makes it that much more difficult. On page 116 of *Real Time II*, after asking what makes a property essential and then rehearsing some doubts about what could distinguish essential from accidental properties, Mellor goes on to say that 'what we need to answer is a different if related question, namely what apparently *changeable* properties are essential to a thing's *survival*, i.e. are such that losing them would be the end of it'. As he says on page 117, so long as these changes of property take place a few at a time, leaving most properties unaltered, it doesn't matter if in the end *all* a thing's properties change. He then goes on to add that 'what keeps any property of *a* unchanged is causation: *a* is *F* at *t´ because* it is *F* at *t*. So for *a* to change from being *G* at *t* to being *G´* at *t´*, causation must link some other facts about it at those two times'.

The curious fact about this account of persisting things lies in the innocent little letter *a* which names the subject of predication, the thing, for the purposes of the explanation. For it might easily seem that *a* is already individuated, its identity settled, prior to Mellor's asking any questions about its unchanging as opposed to its changing properties, and this would make the explanation in terms of factual causation redundant. Clearly Mellor sees *a* as the name of a set of largely unchanging properties - but then how is the bundle *individuated* in the first place? What the ipongo examples were designed to bring out was precisely the fact that our primitive notion of thing identity has to do with transportability in the sense indicated, and that things are individuated in this way prior to our reaching any conclusions about their essential properties. Situations where identity is problematical, where we start out with a *hypothesis* about what appear to be essential properties before we pronounce them as *really* so, as *really* the essential properties of what we then proceed to identify as a definite *something*, must be relatively rare - the 'problem cases' much loved by philosophers. If I am clearing away unwanted rubbish, I hardly wait to go through such an elaborate procedure - I just pile the things in question into the dustbin, or the skip.

The conclusion that follows is that tenseless theory *cannot* accommodate persisting things, things that endure through time. The only road left open for the tenseless theorist is that of perduring rather than enduring things, viz. the thesis that things have temporal parts, with change explained in terms of the temporal parts of a thing having different properties at different times. We shall see how far this is a viable route for the tenseless theorist in the next section.

5 Tenseless Theory, Temporal Parts and 'Back to Quine'

Why should anyone want to say that a thing (say a chair), like an extended event or process such as the performance of a symphony, has temporal parts? Le Poidevin's answer to this question is that it constitutes the only way that tenseless theory can account for change.[23] As he sees it, the choice for the tenseless theorist is between two alternatives, one more acceptable than the other. The first alternative is that objects are wholly located at times, which means - for tenseless theory, at least as *he* understands it - that they can only be instantaneous. The qualification is important here, for as we have seen already, at least one tenseless theorist (Mellor) takes 'objects' (presumably chairs, tables etc.) to be wholly located at each and every moment of their existence. But be that as it may, this first alternative would amount to the prescription that the term 'object' would be reserved for an instantaneous time-slice - i.e. not even a *segment* - of a four dimensional object. The other alternative - the one Le Poidevin prefers - is that a four dimensional object exhibits change by virtue of differences between successive 'temporal parts'. He devotes much space to how this is to be understood as *change* rather than simply a measure of difference along a dimension (like a poker being hot at one end, but only warm at the other). I shall not follow him in that discussion. I shall comment only on how he deals with the objection, in his own words, that 'if objects are really four dimensional entities with temporal parts, then the important distinction between objects and processes vanishes'.[24]   He does not think that his position is threatened by this objection, and after quoting Russell and Quine in support of the thesis that the distinction between things and events vanishes for four dimensional things, simply dismisses it in one paragraph.[25] I do not think it can be dismissed that easily. In fact, what I hope will become clear in what follows is that there are compelling grounds for concluding that the only *consistent* version of the tenseless theory of time is precisely something like the one Quine puts forward in *Word and Object* - but we have to pay a heavy price.[26]

The basic problem lies in making sense what a four dimensional object is, and how it relates to the three dimensional object which, in tenseless discourse, it replaces. But what is a four dimensional object, and what would an example of such an object look like? Let me say at the outset that I have no problem with calling a symphony a four dimensional object - even though there is a distinction to be made, a type-token distinction, between the symphony and *performances* of the symphony (or playings of the CD). Is the symphony *itself* (unperformed) a four dimensional object, or is it only live performances and playings of the CD (etc.) that warrant this title? This is an important question, because how we answer it throws light on whether a particular four dimensional object could, counterfactually, have been in any respects different. If we say that the symphony - the type - is a four dimensional object, then we can say such things as that it might have been played faster if conducted by *A* rather than *B*. We may even properly say that a particular *performance* might have been different - for example, we know that this particular conductor, with this particular orchestra, has played this symphony four or five times, and so we can raise questions about *this* performance, the one we have just heard, and make observations such as 'he could have played the slow movement slightly faster'.   Hence we *can* say that this particular performance might have been different.   This is because of the way a performance owes its identity to the occasion, the orchestra and the conductor.   I could not say, of my headache this morning, that *it* could have been more severe, or more persistent, without meaning that I could have had *a* more severe, or *a* more persistent, headache - even *raising* the question whether it might have been *the same* headache but more severe or longer lasting is just nonsensical.

The major difficulties for the temporal parts theorist come from extending the language of four dimensional objects and temporal parts to tables, chairs, men, women, horses etc.   As Mellor notes, 'when Churchill published an account of his early life, he called it *My Early Life*, not *Early Me*'.[27]   This is not, of course, to deny - in fact it licences - talk of temporal parts of *lives*, as when someone may say 'my childhood was spent in Berlin'.   But we have to note, in passing, another problem faced by people, chairs, tsetse flies etc. that is not faced by symphonies.   One could imagine a performance of a symphony in which, say, the first movement was missed out, or a symphony with its first movement missing because the composer scrapped it, but died before he could re-write it - an incomplete performance in the first case, an incomplete *composition* in the second. Consider the tsetse fly (Le Poidevin's example).   You could cut off spatial parts of it - say, a leg or two - and end up with a mutilated tsetse fly, but a tsetse fly nonetheless.   If, on the other hand, you removed its larva stage, this would abort the whole thing - the demise of its earlier phase does not just remove a *part* of it, but *all* of it.   If 'early me' went out of existence, the amputation would be so radical that it would dispose of 'later me' as well.

If the conclusions of the last section are right, tenseless theory *has* to be committed to a conception of the future as closed.   But if the future is closed, an ontology of persisting things is impossible - impossible because a world in which the future is closed is one where notions of alternative possibility and context transcendence cannot get a foothold.   Such a world would be too much of a plenum to allow for the possibility that my pen, now on my desk, might instead have been in my pocket.   We normally make some sort of distinction between the chair itself on the one hand, and what has happened to it, what *might* in the future happen to it, and what (counterfactually) *might have happened* to it had circumstances been different, on the other.   But in the case of the corresponding four dimensional object, this distinction goes - there is no longer a chair, and so the distinction between it and whatever might happen to it cannot be invoked.   This surely threatens the very idea of a four dimensional object. What we need is some kind of analogue, in the language of four dimensional things, for just those central features of a three dimensional thing (and its identity through time) referred to above, which have to do with the distinction between the object itself, and what happens to it (or *might* happen to it, or *could have* happened to it in counterfactually different circumstances). Where is such an analogue to come from? No answer to this question is forthcoming.

If I say, 'my car could have been different - it could have been a Mercedes', I would be taken to mean 'I might have had a Mercedes instead of a Ford, had circumstances been different'. What I could *not* mean is that my Ford Escort - that particular car - might have been a Mercedes. (I *could* have had a Mercedes, had I been richer.) Equally, there is a place for the statement 'my Ford Escort could have been different' - for example, it could have been in better condition if I had treated it against rust. The two kinds of statement are *distinguishable* when applied to three dimensional things. But they are not in the case of four dimensional objects - we just cannot make the distinction.

What underlies this failure is the fact that, once the ordinary conception of a three dimensional object capable of undergoing changes has gone, there is no longer any basis for identifying and reidentifying 'objective particulars' in Strawson's sense.[28] The upshot is that the tenseless theorist *has* to be committed to a conception of the universe according to which there are, simply, properties and qualities at times and places in a closed, four dimensional continuum of three spatial and one temporal dimension. There would be, were such a umiverse conceivable anyway, *no basis at all* for distinguishing one thing from another, and therefore no basis for assigning a region of the 'four-space' to one 'object' as opposed to another 'object'. How we individuated so-called 'four dimensional things' would then be a purely arbitrary matter - a four dimensional object would be whatever we *say* it is. We can regard (say) a bicycle from its construction to its final demise as a single four dimensional object. But why should we? Surely to do so is to make the three dimensional object that 'persists through time' ontologically prior to the four dimensional object, thus reintroducing by the back door what the 'temporal parts' theorist has expelled (or tried to expel) from the front. Merely *loosening*, rather than breaking completely, the primordial link with three dimensional things, on the other hand, licences some bizarre possibilities. For example, a segment of the history of a chair together with a segment of the history of a table in the 16th century could equally constitute a four-dimensional object - or, perhaps *slightly* less bizarre, a segment of the chair's history and a segment of the history of a gatepost (where the two segments are contemporary) could constitute a four dimensional object. Breaking the link completely - which is, logically, what the tenseless theorist must do - leaves us with a total plenum (or total void, it comes to the same thing) in which no distinctions are possible between one thing and another.

What inevitably follows from this is that the only tenseless theory that 'works' - if 'works' is the right word here - is one like Quine's, as propounded in *Word and Object*.[29] The central tenet of this theory is that the distinction between things and events is obliterated. But it may be even worse than that - if the distinction *is* obliterated, do we not lose both sides of the distinction?

Notes and References

[1] See J.M.E. McTaggart, *The Nature of Existence*, Cambridge 1921, volume 2 chapter 33; also 'The Unreality of Time', in J.M.E. McTaggart, *Philosophical Studies*, London 1934, (originally published in *Mind* 1908).

[2] See C.D. Broad, *An Examination of McTaggart's Philosophy*, Cambridge 1938, quoted in R.M. Gale (ed.), *The Philosophy of Time*, London 1968, pp. 117-142; G.E. Moore, *The Commonplace Book of G.E. Moore*, Casimir Lewy (ed.), London 1962, pp. 404-5; and David Pears, 'Time, Truth and Inference', in Antony Flew (ed.), *Essays in Conceptual Analysis*, London 1966, pp. 228-252.

[3] Andros Loizou, *The Reality of Time* (hereafter *RT*).

[4] McTaggart, 1921, volume 2, pp. 9-10.

[5] Ibid., p. 10 (my italics).

[6] C.D. Broad, in R.M. Gale (ed.), *The Philosophy of Time*, pp. 118-119.

[7] See R.M. Gale, *The Language of Time*, London 1968, chapter 6.

[8] See *RT* chapter 3 passim, but especially section 3.1.

[9] McTaggart, 1921, volume 2 pp. 10-11.

[10] See p. 126, footnote. This and other references to the *Mind* 1908 article are from McTaggart, 1934. See note 1 above.

[11] McTaggart, 1921, volume 2, p. 11.

[12] Ibid. pp. 9-31, and 'The Unreality of Time' passim.

[13] In the 1908 article, he introduces the C series thus: 'There is a series - a series of the permanent relations to one another *of those realities which in time are events* - and it is the combination of this series with the A determinations which gives us time. But this other series - let us call it the C series - is not temporal, for it involves no change, but only an order' (McTaggart, 1934, p. 116, my italics).

[14] McTaggart, 1921, p. 13.

[15] McTaggart, 1934, p. 114.

[16] R. Le Poidevin, *Change, Cause and Contradiction*, London 1991, p. 130.

[17] The term 'B-statements' or 'B-sentences' is used by D.H. Mellor in *Real Time II*, London 1998, to state such 'B-facts' as that 'Jim *runs* at *t*', where 'runs' is read tenselessly.

[18] L. Wittgenstein, *Philosophical Investigations*, Oxford, 1963, part I para. 270.

[19] R. Le Poidevin, *op. cit.*.

[20] D.H. Mellor, *op. cit.*, pp. 85-6.

[21] But there are cases where a particular thing, if we construe the notion of a thing widely enough, owes its identity to an event. A crater would be an obvious example - it owes its identity not to the meteor as such, but to the *falling* of the meteor.

[22] See P.F. Strawson, *The Bounds of Sense*, London, 1975, pp. 224-33, p. 122*ff.*, and elswhere.

[23] R. Le Poidevin, *op. cit.*, see especially pp. 9-10, pp. 58-75.

[24] Ibid., p. 65.

[25] Ibid., pp. 65-6.

[26] W.V.O. Quine, *Word and Object*, Cambridge Massachusetts, 1960, pp. 170-176.

[27] Mellor, *Real Time II*, p. 86.

[28] See P.F. Strawson, *Individuals: An Essay in Descriptive Metaphysics*, London, 1961, chapter 1.

[29] Quine, *Word and Object*, p. 171.

2 The Dynamic of Time

1 Introduction

The idea that events 'take place'; the idea that situations 'come to pass'; the idea that time 'passes', 'flows' or 'elapses' - all these, and their like, are taken for granted by ordinary folk innocent of philosophy. Philosophical reflection upon them throws up all kinds of problems and paradoxes, such as St Augustine's problem with the extent of the present, and McTaggart's attempt to prove that time is unreal.

The problem posed by the need to give some kind of account of these aspects of time is one of the hardest, if not *the* hardest, that philosophy (more specifically, metaphysics) has to face. The problem is conveniently bypassed by tenseless theorists. However, tenseless theorists are clearly dependent on McTaggart's A series/B series distinction, both in advancing their own view and in their polemic against tensed theories. The tenseless theory is flawed for the reasons given in chapter one, which include among them the fact that it leans heavily on the language of McTaggart. (There are *tensed* theories which lean equally heavily on McTaggart, and so they too will be flawed for that reason - but more of this later.) I claimed in the last chapter that the main error bequeathed by McTaggart is the gross caricature he gives us of the passage of time, the central feature of this being his insistence that the tense or A determinations of events are properties or qualities, and hence that the passage of time consists in events successively losing and acquiring such properties or qualities.[1] In this chapter, I shall address the issue of how we can free ourselves from this picture and begin to redraw the metaphysics of time.

Before attempting this, we need to remind ourselves about what is wrong with the picture. There are two things, closely related. The first is that there is the temptation to think of changes of tense determination as fortuitous, accidental and external to an event. If 'becoming past' is thought of as like 'turning blue' - or, perhaps more illuminatingly, like 'being sold to X' - then 'becoming past' does come to look external and fortuitous, and certainly leaves the event essentially the same before and after the change, to echo McTaggart's words. In fact, the second simile - where 'becoming past' is like 'being sold to X' - fits the situation better than the first. The second and closely related thing is that the picture fails to do justice to the necessity by which, for example, a present event *must* become past. So if we are to draw an alternative picture, that picture must show change of tense determination to be both necessary and an internal aspect of time.

It is because McTaggart's account of temporal passage is such a gross parody that he is able to mount his 'proof' that time (or more specifically the A series) is unreal, and so I begin with some comments on this. The proof only works - if it works *at all* - on the assumption that we have an event *e* to which contradictory predicates P, N, F ('-is past', '-is present', '-is future') apply, i.e. so that the conjunction of the three propositions P$e\rightarrow$ ~[Ne & Fe], N$e\rightarrow$ ~[Pe & Fe] and F$e\rightarrow$ ~[Pe & Ne] with the proposition [Pe & Ne & Fe] is contradictory. What is immediately clear from this formal notation is that we have an antecedently individuated event *e* to which a series of predicates apply *ab extra*. To repeat what I said above, redrawing the metaphysics of time requires that we come to see the changes of tense-determination which an event undergoes as both an internal aspect of time and a necessary aspect. If we are to do this, then the notation we use has to be recast accordingly. I shall have more to say on this in what follows.

We need, first, to remind ourselves about the paradoxical nature of McTaggart's account of the A series and the B series. The discussion in section 2 of chapter one can be summarised thus: McTaggart gives us, on the one hand, a B series, and on the other, an A series that is in reality nothing more than a B series, artificially set in motion and scrolled past the observer situated beside it and looking on, providing us in this way with an image of moving time in which nothing but the most superficial aspects of the distinction between past, present and future find expression. The 'transition' from *e* being future to *e* being present, and from *e* being present to *e* being past, must be other than what we get from this picture, in ways yet to be determined. If we look at how these transitions feature in ordinary experience, we may perhaps be guided towards a new understanding. I turn to this in the next section.

2 Experience and the Present: a First Look

The problem with trying to look afresh at experience lies in the difficulties that lie in the way of divesting ourselves of received prejudices, metaphors that were once living but are now so dead that they wear the garb of literal truth, and - if this *is* a further category - the sedimented theories of yesterday. We sometimes hear, for example, that 'no extended event is ever *strictly speaking* present' (although it may be 'speciously' so), 'because the present cannot be other than an instant'. Again, we sometimes hear it said that 'we can never be aware of the past itself, only a representation or trace of it in the present'. Some of these common sayings are in fact wrong -

both the aforementioned fall into this category, as will emerge presently. The attempt to describe experience afresh is a twofold task. On the one hand, we have to direct our inquiries at experience itself - experience in general, or particular kinds of case - and on the other, we have to be continually vigilant about the actual *words* we use in characterising experience.   But the very possibility of vigilance in this matter clearly implies that we can come to know when our efforts have proved successful, and this in turn implies that there are at least *some* parts of our temporal language that we can trust.

The way of proceeding, then, should be this: we look at experience along with ordinary, pre-theoretical ways of talking about events and time, about past, present and future, about earlier and later, and so on. But we are to suspend judgement about general propositions like 'no extended event is ever strictly speaking present'. If, as a result of our vigilance, we find that one or another of these general propositions is untenable, then it must be rejected.

It will be convenient to begin with St Augustine, who in book 11 of his *Confessions*, raises his well-known problem about the extent of the present, and after some considerable reflection provides us with an answer - but an answer which is not as satisfactory to us his modern readers as it was to him. But it is nonetheless instructive for us to consider both the problem as he stated it and his proposed solution.

Without going into detailed textual exegesis, there are two aspects to his account of the present: what we might call, respectively, the 'extent' aspect and the 'existence' aspect. The 'extent' aspect can be summarised in the following way. Let us suppose someone asks me 'what are you doing now?' and I reply 'I am working on my book'. My inquirer is not satisfied with my answer, and so takes his questioning further and asks me to be more precise - and so I say to him 'I am working on chapter two'. He is not satisfied with this either, and so I tell him I am working on the second section of chapter two. But he persists in his quest for what he considers to be greater accuracy, and so I tell him I am writing a paragraph about St Augustine. He is still unsatisfied with my answer, so I tell him I am writing such-and-such a word. He presses me further, and so I narrow my reply down to writing such-and-such a *letter* - and so on, indefinitely.   His purpose is to get me to accept that no event having temporal extent is ever 'strictly speaking' present, and he would have me replace a statement like 'I am writing such-and-such' with the conjunction 'I have been writing such-and-such and will be writing such-and-such'.

The 'existence' aspect of Augustine's account is typically expressed in such statements as 'the past is now no longer' and 'the future is now not yet'; and his questioning takes the form of asking, how can I measure time if the past is now no longer, the future is now not yet, and the present is extensionless? I cannot measure past time because it is no longer there to be measured, I cannot measure the future because it is not yet in existence, and I cannot measure time while it is passing in the present because the present lacks extension. And yet, it is important to notice that while Augustine is thus concerned *theoretically* with the extent of the present, while he is puzzled as to how it is *possible* for us to measure time, he nonetheless holds on to the fact that we *do* measure it. But what, in the end do I measure? What is Augustine's answer?

After what looks like a detour in which Augustine asks whether time might not be motion, he eventually arrives at the proposition that time must be an extension of something, but he knows not what. Then he moves on to the proposition that we measure not time itself, but something in the mind that remains fixed - and thence to the *identification* of the extension that is measured as time with the *distentio animi* or extension in the mind by which it is measured. It is not time itself that is measured, for this does not exist in a form in which it can be measured. Time itself is not long or short, but instead we reckon with 'a long memory of the past' and 'a long expectation of the future'. As I pointed out in *RT*, Augustine in effect claims that there is a kind of ontological deficiency in time itself, and that by conjuring up 'an extension in the mind itself' *we* are somehow able to make up for this deficiency.[2] Now while it is easy enough to see why this answer is unsatisfactory - if time itself is ontologically incomplete, it is hard to see how an extension in the mind can make good the deficiency - what does need pointing out is how we are the inheritors of Augustine's thesis that the present is 'strictly speaking' extensionless. This is borne out by the tell-tale phrase 'the *specious* present' - the present of experience is 'specious' in that it is never 'really' extended, although it gives the deceptive appearance of being so.

In response to Augustine's problem, I shall maintain that we do not need to resort to 'extensions in the mind' in order to account for extended present experience *per se*, and that it is possible to speak of extended events as present *in the strict sense*. This will entail a radical questioning of the nature of our experience of time and of the way we describe it. At this stage in the inquiry I shall focus exclusively on our passive experiencing of events and processes, our experience as perceivers rather than agents. In this, I shall draw, in spirit if not in the actual detail, on Husserl's *Phenomenology of Internal Time-Consciousness* and commentators on this work. But first, some preliminary remarks.

Let us begin with how we speak about present events - as when we say, for example, 'I am writing a letter to *A*' or 'I am speaking on the telephone to *B*'. In both these examples, it can be legitimately inferred that I have begun writing to *A*, or talking to *B*, and have not yet stopped. If we followed Augustine, we would have to convert these statements into 'I have been and will continue to be writing a letter to *A*', 'I have been and will continue to be speaking to *B* on the telephone'. The use of the continuous present tense, 'X is φ-ing', is just not a legitimate *sui generis* way of speaking at all. Augustine's logic seems impeccable - how can any extended event, which admittedly we *speak* of using the form 'X is φ-ing', really be present, if all its parts are 'in the past' or 'in the future'? The basis of Augustine's argument is the undeniable principle 'if X is present, it is not past or future; if X is past, it is not present or future; and if X is future, it is not past or present'. How, then, can we speak of the present as extended without contradiction?

Let us suppose that I am writing a letter, and therefore that I have begun and not yet stopped. Let X be the whole event from start to finish, and let p be a past part of X, and f a future part. Thus while I am, as we say, writing a letter to John, there is something p I have done (e.g. writing my address at the top of the page), and something else f that I have not yet done (e.g. signing the letter). Both p and f are *parts* of X, which *as a whole* we speak of using the continuous present tense ('I *am writing* a letter to John'), yet p is past and f is future. Are we forced to deny that X is 'really' present, because p and f are past and future respectively? Or do we have to deny that p is 'really' past and f is 'really' future if we are to keep hold of X as present? Surprisingly, the answer is No to both questions.

In saying that X is present, we do not have to deny either that p is past or that f is future. The statements 'X is present', 'p is past' and 'f is future' are not in contradiction, because the subject of predication is in each case different. They would only be contradictory if we added as a premise the principle 'if X is present, then all its parts are present'. The analogous principle for space would be 'if B is here, then all its parts or contents are here' - if the box is here, and the box contains marbles, then all the marbles are here. But the principle 'if X is present then all its parts are present' is not presupposed by ordinary usage, which allows us to say that X is present, p and f are parts of X, p is past and f is future without contradiction. If ordinary speech is to be our guide, then we can say, without contradiction, that an extended present event will have past and future parts. A past or a future event will, by contrast, have all its parts respectively past or future.

These facts of ordinary speech are no more than a clue. I would not want to claim that by themselves they constitute any kind of justification of the principle that an extended present event must have past and future parts or phases. Much more needs to be said.

I begin by drawing attention, again, to the terms 'is now no longer' and 'is now not yet' as applied by Augustine to the past and the future respectively. The privative nature of these terms makes an alternative view to Augustine's very difficult to articulate. Not all experiences of the past are privative, least of all those of the immediate past. Suppose I am listening to a piece of music, and that I am in the middle of a certain phrase - say, the opening phrase of Beethoven's fourth piano concerto. The opening chord, played softly on the piano, is indeed past, and I could describe it as a past part of the phrase I am now listening to. The chord is repeated three times, then followed by a succession of other chords which bring the phrase to a close. At the end of the phrase, or half way through it, is it plausible to speak of the opening chord in privative terms? Is it not rather experienced as initiating the phrase, and therefore as the foundation that is enriched by the parts of the phrase that follow? The point could be made by reference to a cricketer who has just hit the ball and is starting to run to the other end of the pitch. Do we, the spectators, as we watch him running, experience the earlier moment when he hit the ball as a privation? Is it not more natural to characterise it in more positive terms, e.g. as something full of promise?

To follow these rhetorical questions through is to move towards something radically different from the received view. It is to move towards the notion of a past - or at least, a *just*-past - that is immediately and directly given. This is the view taken by Husserl.[3] In what follows, I hope to show that it is the *only* possible view, once we become aware of the deficiencies of the alternatives. Before recommending this as a conclusion, I shall summarise the difficulties lying in the way of the received alternatives. I shall consider two - the account of our awareness of an extended present in terms of present mental contents explicitly endorsed by empiricists, and the account known as 'the specious present theory', which is in reality no more than an operationalised version of the empiricist thesis anyway.

Let us begin with the pure empiricist thesis. The past itself (albeit only the just-past) is gone - 'is no longer' - and so what forms the content of my temporal awareness, what gives me the sense of temporal 'thickness', is the persistence in my mind of a present mental content of some sort. That is the view under consideration.

The account does not work, on either of the two ways there are of conceiving the persistence of a present content. If it is conceived as the persistence of all elements of the content simultaneously, it remains wholly mysterious how they can be perceived *successively*. If the extended event were a phrase of music, for example, the simultaneous persistence of all elements of the mental content could only give me the notes of the melody *all at once* rather than in succession. (There have even been attempts to make a virtue of the perception all at once of rapid succession - e.g the rapid movement of a cricket bat hitting the ball, which (it is claimed) is perceived as a simultaneous arc embacing the whole movement; as if this could function as some kind of explanation of the perception of temporal extent generally.) If, on the other hand, the mental content is conceived as a succession of representations - so that the representations of the notes of a tune occur in the mind one after the other - the original problem calling for explanation recurs. The past parts of the representation are gone, and so we have to invoke a *second* level of representations which persist in the present.

Let us turn now to the theory of 'the specious present' - so named because of the presupposition (whether acknowledged or not) that the present was still 'in reality' extensionless.[4] The theory was meant to reveal the length of time that would be known directly, in immediate experience, without the help of memory, or - in another formulation - that length of time (taken to be a maximum) which could be apprehended in an unbroken act, or span, of attention. It was usually, though not exclusively, understood to involve only the immediate past and not the future.[5] Psychologists performed experiments in order to measure this. A typical experiment was one where the subject was presented with a series of unfamiliar sounds, and then asked to reproduce them correctly. The length was increased until the subject began to make mistakes. The critical length - the maximum length that the subject could reproduce without making mistakes - was deemed to be the specious present.

There are many questions that can be asked about this theory, but here I shall confine myself to one point.[6] The theory relies exclusively on the notion of an unbroken attentive span. The unbroken attentive span which constituted the specious present was deemed to stretch into the past for five or six seconds, with a sharp cut-off. Clearly the content of this attentive span would vary from moment to moment, and just as fresh events entered it, so the immediately preceding events would leave it, abruptly, at its trailing edge. And my point is about the very unity of this attentive span, carried forward from moment to moment. The only temporal wholes I could apprehend would be those which fell within the six second span of attention. Events before that would be continually falling off the edge, as new events entered awareness. On the specious present theory, I could never properly apprehend a phrase of music. To illustrate this, consider that I am presented with a temporal object - a tune, say, represented by ABCDEFG. Let us suppose further that my trailing attentive span can accommodate three notes at a time. The content of the first saturated attentive span will then consist of ABC. This will then be replaced by BCD, then by CDE, then by DEF, and then by EFG. On this, I can *never* perceive ABCDEFG as such. The unity of this ever-moving attentive span has been bought at the expense of the unity of the temporal object. This is surely too high a price to pay.

If my specious present is such as to accommodate only three notes at a time, then the contents of my mind from moment to moment will be represented by ABC shading into BCD, BCD shading into CDE, and so on. In the diagram below, the right-hand vertical line represents the present instant (N), and the two vertical lines taken together represent that interval which is the specious present (SP). The contents of the specious present will therefore be whatever falls between the two lines. The content of my awareness of the tune include first nothing (the whole melody is future), then A, then AB, then ABC, then BCD, and so on, ending with the final note G and then nothing, as the whole melody goes into the past.

PAST	SP	N	FUTURE
			ABCDEFG
	A		BCDEFG
	AB		CDEFG
	ABC		DEFG
A	BCD		EFG
AB	CDE		FG
ABC	DEF		G
ABCD	EFG		
ABCDE	FG		
ABCDEF	G		
ABCDEFG			

Figure 1

What operates here is a very crude notion of the present. The specious present is like a container in which there is a compresence of elements with a saturation point. If anything deserves the title of 'a crude metaphysics of presence', it is this. The transitions from ABC, to BCD, to CDE etc. are no more than a list of successive *contents* of awareness, with no element of meaning - the meaning of the temporal object itself, the melody, is lost. We gain a non-instantaneous present, a 'unity of content' of sorts - but at a high price. It is a subjective unity only, with no regard for the melody as an objectivity. While it does not *explicitly* refer to present mental contents, traces etc., the theory nonetheless represents a desperate attempt to hang on to these ideas.

What both the specious present theory and the empiricist theory generally give us, well represented in the diagram, is a crude, self-enclosed and hence non-referential notion of experiential content. Bound up with this is a notion of 'the world' as equally self-enclosed and perspectiveless, depleted of all significance and meaning. On the one hand, we have a 'consciousness' or 'subjective experience' made up of successive contents - A, displaced by AB, then by ABC, then by BCD (and so on), which fails to adumbrate a perspective on anything outside it. On the other hand, we have an 'objective world' incapable of sustaining any significance, composed of isolated elements - A, then B, then C etc. - which carry no significance for one another beyond the bare facts of contiguity, succession and other such abstract determinations.

3 The Present and the Just-Past

The problems generated by these theories point to the need to acknowledge that the past, or at least the just-past, is directly given in experience.[7] I shall now address this directly.

I begin with a point of terminology. In place of the term 'tense determinations', I shall henceforth speak of 'tense profiles' or 'temporal profiles'. The rationâle is this. Just as we speak of different perspectives on or elevations of, say, a house, so we can speak of different perspectives or profiles of an event - e.g. the event as present, just-past, a minute ago etc.. As will become clear presently, the terms 'tense profile' and 'temporal profile' are better suited for the purpose than 'tense determination'.

We need to leave behind, once and for all, the idea that the passage of time consists in events successively acquiring or losing pastness and presentness as if the latter were *properties* or *qualities*. What I hope will become clear is that speaking of tense or temporal 'profiles' rather than 'determinations' helps us to leave such ideas behind. For whereas speaking of 'the changing tense determinations of events' carries suggestions that this and then that quality or property is *predicated of* events, the language of profiles and perspectives certainly does not. It is no more natural to think of a tense or temporal profile as predicated of an antecedently individuated event than it is to think of the perspective from which the house is seen as predicated of the house. A view of the house gives us *the house* - the house as seen from the corner of the street, let us say - and not the house together with some extraneous attribute. A tense profile can thus be seen as a perspective in which the event shows itself, and what is given in that perspective *is* the event, not the event plus some quality added to it from the outside. The tenseless theorists are therefore right in the scorn they pour on the idea of tensed *properties*, but wrong in so far as they prioritise a tenseless time and relegate tense to merely relating the speaker to the context of his utterances and 'beliefs.'

What it is important to grasp is this: just as there is no such entity as 'the house from no perspective at all', so there is no such entity as as an event without a tense or temporal profile. And it is inescapably true that I can 'run' the same perspective on the house - e.g. the view of it from my study window - as many times as I like, but that I cannot 'run' again the same temporal profile of a past event. What I can do is recall it again and again; but these later occasions of recall are different tense profiles. Even recalling the occasion of my first or original recall does not give me 'that temporal profile again' - it gives me a complex profile, i.e. a profile of an earlier profile.

If we think in terms of temporal or tense 'profiles' instead of tense 'determinations', it becomes easier to accommodate the idea that the just-past is directly given. For Husserl, as an event runs its course, it does so through a continuous cascade of temporal profiles to which he gives the name 'abschattungen', which literally means shadows cast off by something. While in the midst of a phrase of music, I hear a particular phase of the tune, a note - let us call it B - in the context of having already heard, 'a moment ago', the note C, and with the expectation that it will be followed by another note, say A. Again B itself has some duration, and while I am *hearing* B, part of it has elapsed and part of it is to come. *Unless we can look upon the elapsed phase and the phase to come of the note B as parts of an extended present experience of B, then all talk of 'the present' becomes unintelligible.* Our temporal experience, then, has to be described in such a way that a succession of events or phases is immediately given *as such* - so that I am aware of the immediate past and the immediate future not via 'images', 'traces' or other proxy items, but directly.

In other words, the immediate past and the immediate future *themselves* are part of the content of any experience of a present event. Furthermore, as I hope to show in what follows, only by speaking of events in this way is it possible to give an account of the *consciousness* of the temporal object (the event, process or whatever) in a way that preserves the unity of the temporal object itself. The theory of the specious present, as we have seen, was unable to do this, precisely because of its naive, simplistic notion of what temporal presence consists in - a 'presence' where the so-called 'unity' of the act of consciousness is bought at the expense of the unity of the temporal object. I said enough in section 2, in general terms, about these limited (and limiting) notions of the present. But there is some debate about whether Husserl himself may not have fallen prey to these very notions which he is at pains to differentiate himself from. To follow these debates in any detail would take me far beyond my aims in this essay. I shall nonetheless make one or two general points, with the twofold aim of clarifying what Husserl is trying to do, and then offering my own parallel account of our experience of time. In this I shall be concerned with doing justice to the unity of the temporal object itself, whilst at the same time taking the resulting account as far away as is possible from 'psychologistic' interpretations.[8]

Perhaps a good place to begin is the theory of time-consciousness which Husserl is at pains to reject, a theory he ascribes to Brentano.[9] The theory as Husserl presents it - the qualification is important - begins with the claim that in order to reach a correct understanding of temporal experience, more specifically the experience of succession, we need somehow to steer between two erroneous pictures. The just-past phases of an event such as a tune cannot be simply present to consciousness in the same way as the present note, for then we would hear not the notes of the tune in succession but a jumble of notes simultaneously. On the other hand, they cannot simply *disappear* from consciousness, for then we would not experience succession. The past must, somehow, be held together with the present - and Brentano's account of how this is possible is, according to Husserl, the intermediate one of postulating the existence of 'primordial associations' which, when the past notes die away, 'represent' them in their absence by attaching themselves to the present phase and enriching it with a temporal character.

The parallel with Augustine is clear. We do not have the past note itself, as this is 'now no longer', but in its place we have something in the mind, a mental reptresentation. In contrast to this, Husserl wants to say that the past note or past phase is directly given *as such*, and not via a proxy such as an image or representation - the past is there, joined to the present, and despite being 'no longer', is nonetheless present *in person* but as itself, namely as *past*. This may seem a contradictory position. If it does, it is because we are still held captive by the notion that it is *only* by means of 'representations' that the 'now-no-longer', even the *just*-past, can show itself to consciousness. That the past is nonetheless present in person is what Husserl wants to say, and it is this that he intends when he speaks of 'retention' as a mode of 'originary givenness'.[10]

Husserl illustrates his theory in his famous 'diagram of time', which I reproduce below in Merleau-Ponty's slightly modified form:[11]

Past   A   B   C   Future

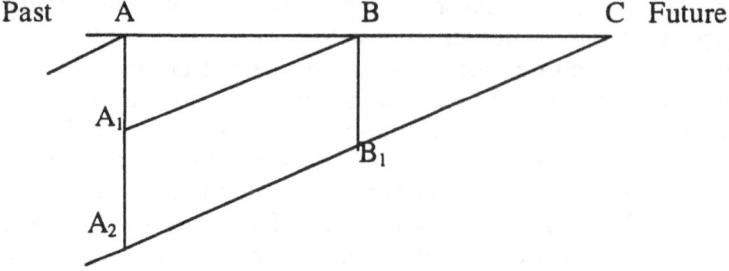

Figure 2

A, B and C are (let us say) three successive notes of a melody, or three successive phases of a single enduring sound. We ordinarily think of the succession A-B-C as a case of first A sounding, then B sounding, then C, in a way that corresponds to the following picture, with the present 'moving on' from A, then to B, then to C:

Past   A_____B_____C   Future

Figure 3

Clearly from the standpoint of C, i.e. when C is present, A and B as simply past stand in need of some 'synthesis' by which they can be related to C as the earlier notes of the melody. The traditional 'epistemological' answer has been in terms of images and representations. Husserl's main purpose is to show up the epistemological answer as inadequate, and to recommend an alternative. The clue lies in Husserl's 'diagram of time' and how it goes beyond the picture suggested by figure 3. We can view Husserl's alternative as an account of the 'going on', 'taking place' or 'coming-to-pass' of a single event, series of events or process which aims to get beyond the parody of time that analytical philosophers have inherited from McTaggart's A series. The measure of Husserl's success must lie, I shall suggest, in how far he gives us a convincing account of the coming-to-pass of an event as a nexus of *intrinsic* modifications, and whether this account takes us beyond the view that the changing tense-profiles of an event are changes *happening to* the event, rather than modes or perspectives in which the event manifests itself, puts itself forth or 'appears'. To this question I now turn.

The important features of Husserl's diagram (figure 2) lie in the addition of vertical and diagonal lines to what would otherwise be the single horizontal line of figure 3. There are vertical lines linking A, A_1 and A_2 and B and B_1 respectively; and there are diagonals linking, first, B and A_1, and then C, B_1 and A_2. First, A sounds. Then, when B sounds, it is joined to the elapsed A as A_1, as just-past, as a 'retention'. A_1 is, in other words, part of the extended event which is 'the melody I am listening to now' - but it is nonetheless just-past, and directly or 'originarily' given as such. Essentially, the question Husserl is addressing is this: how can we conceive 'A followed by B' in a way that does *not* invite the thought 'A, and then *no longer* A but B' and what goes with this thought, namely that since A 'is no longer', what is sustained along with B is a mere *representation* of A? Husserl's answer is: only by thinking in terms of the retention of the just-past as a mode of originary givenness. In the terms I used extensively in *RT*, 'A and then B', or 'future *then* present *then* past', gives us only half the picture; we need also to think in terms of a future *with* present *with* past principle. Hence in our example, A gives way to, or shades into, B-with-A_1.[12] The 'with' is meant to convey the thought that it is A *itself* that is the constituent of the conjunction, not a mere *representation* of A. With the advent of C, B and A are now held as its antecedents in retention, i.e. non-representationally. Just as A gives way to B-with-A_1, so B-with-A_1 gives way to C-with-(B_1-with-A_2). These conjunctive phrases (B-with-A_1 and so on) correspond to the diagonal lines in Husserl's diagram. More needs to be said about these diagonal lines, and in particular the significance of the change of suffix as A is modified first into A_1, then into A_2, then into A_3 and so on. What, it might be asked, is gained by this procedure? The answer is complex, and leads naturally to the task of redrawing the metaphysics of time. I turn to this in the section that follows.

4 The Micro-Structure of 'Going On'

The aim, let us recall, is to get beyond the linear picture given in figure 3, which we can without distortion view as a diagramatic representation of McTaggart's A series, with presentness moving to later and later terms: first A is present, then as A loses presentness and acquires the property of pastness, B loses the property of futurity and acquires the property of presentness - the point being that the changing tense determinations of A and B are viewed as changes of property or quality which happen to the event *ab extra*. The upshot of this picture is, as we saw in chapter one, that tense-determinations as such come to seem more and more superfluous, evanescent, unreal, even contradictory as McTaggart believes - and this leads inexorably down the only roads left open, namely either to the *toto caelo* unreality of time, or to the prioritisation of McTaggart's B series and the tenseless theory of time.

Insistence on the suffixes in Husserl's diagram marks a decision - let us initially put it no stronger than that - to view the cascade of tense profiles that manifest an event as *inseparable from* the event, but not as properties or qualities *predicated of* the event as antecedently individuated. To think in terms of events as antecedently individuated, i.e. prior to their tense profiles, is to be the prisoner of a dogma. An event is not a 'something else' that stands behind its succession of tense profiles, not a kind of reality of which the tense profiles are merely the appearances - these profiles simply *are* the event, without remainder. The order in which an event manifests through its tense profiles is a necessary one. In Husserl's terms, A is modified into A_1, A_1 into A_2, A_2 into A_3, and so on. And there is no 'ideal' A, A_T let us call it, which gives us A from no perspective at all, or (what is perhaps the same thing) A from all possible perspectives at once. To believe in the primacy of A_T over A, A_1, A_2, etc. is precisely to believe in the logical fiction that fuels tenseless theory, i.e. to believe that there is an A_T in the first place which A, A_1, A_2, etc. share as their underlying common essence. It is *precisely* this A_T which is a logical fiction. There is no such thing as A *simpliciter*, i.e. A from no perspective at all, despite Husserl's use of A without a suffix in the case of the primal impression or original 'given'. In what follows, I shall modify Husserl's notation in a way that I hope brings this out. Following the notation I used in *RT*, I shall write

A_0 instead of A, and negative indices for future tense profiles or 'protentions'. (Husserl never gives a notation for protentions, only for retentions.) An event A in the more distant future will have the tense profile A_{-5}, then A_{-4}, then A_{-3}, then A_{-2}, and finally, when it is just-future, A_{-1}. But A_{-5}, A_{-4} etc. should *not* be thought of as definite referring expressions uniquely designating 'events in the future' as if the future were ontologically closed, for the reasons given in section 3 of chapter one. More needs to be said about the sense of the future which Husserl attempts to grasp through 'protention', a sense more primitive than is given by such notions as 'expectation', 'prediction' and so on. I shall return to this presently.

First we need to recall the aim set out at the start of section 2 of this chapter, which, stated in general terms, involved interrogating experience in a way that would unearth erroneous metaphysical assumptions and thereby 'free the ground' for a redrawing of the metaphysics of time. It would be useful to take stock of how far we have gone down that road, before going any further. I turn to this task in what follows.

First: *contra* Augustine, we do not have to think of the past in a negative, privative way as merely 'the no-longer'. When we think of something as having just begun - for example, a favourite piece of music, or the path of a cricket ball which promises to score a six - we are certainly not thinking of the just-past as a privation or 'absence', but more as a solid *foundation* for what 'promises' to follow. Second: we should see this in conjunction with the common use of the continuous present tense. We say 'Peter is writing a letter to John', from which it follows that he has begun writing the letter and not yet stopped; we might then go on to say 'he has written that date', and also 'he has yet to sign the letter'. When we make the last two statements, past and future tensed respectively, the past and future events they describe are *parts* of the present event described in the statement 'Peter is writing a letter to John'. In short, ordinary speech seems to *accommodate* the idea of an extended present event having past and future phases or parts, *without contradiction*. We certainly do not consider the original statement 'Peter is writing a letter to John' to be in need of any *reductive* analysis. Third: we have direct access to the immediate past *itself* without needing to invoke images or representations of any kind, in what Husserl calls 'retention'. We are continually aware, when we hear the opening bars of a piece of music, of the first few notes let us say, without having deliberately and reflectively to *recall* those early notes. Access to the immediate past is thus *not* via any kind of 'representation' - it is direct.

What I see the foregoing as leading towards, and what I hope to complete in the rest of this chapter, is at least a *preliminary* account of what, essentially, an event is, and how events differ from 'things'. We often speak of events as if they were quasi-things. We use nominalisations such as 'John and Mary's wedding' forgetting that they are derived from

'John and Mary got married (or are in process of doing so, or will be doing so)'. And we often talk of an event, e.g. 'the opening phrase of Beethoven's Ninth Symphony', in abstraction from the fact that the *event* is the actual *taking-place*, in a performance, of that phrase. (When we talk abstractly in this way, we may be referring simply to the written score - but the score is merely the set of instructions for *producing* the event, i.e. the performance itself.) An event is, in short, the *taking-place* or the *going-on* of something. It is a necessary part of the project of redrawing the metaphysics of time that we examine the *micro-structure* of such 'going-on' or 'taking-place'. The language of tense profiles should be seen as an essential aid to achieving this.

Husserl's work on time and time-consciousness, whilst having other aims, does aim to provide some account of this micro-structure, and this is the aspect I shall most seek to exploit. There are other points where what Husserl says touches on themes I pursued in *RT* and on which I shall have something to say in this essay - principally his discussion of the 'constitutive flux' in the later sections of the lectures on internal time-consciousness. I return to this aspect later.

Events and processes 'go on' or 'take place'. The object of the present inquiry is to develop an account of the 'going on' of an event (I am thinking of non-instantaneous events here) which gets beyond the simple idea that this 'going on' consists in the event's transition, *already constituted*, from a region called 'the future' via 'the present' and thence to its final destination, the region called 'the past' - or, what is the same thing, the event's having first the property of futurity, then exchanging that property for that of presentness, and then exchanging its presentness for the property of pastness. To say that A is first future, then present and then past, and then to say the same of any non-instantaneous part of A, is not to offer anything like an *account* of the 'going on' or 'taking place' of A. To speak in this way is to remain fixated at the level of what Bosanquet calls, admittedly in another context, 'theories of the first look'. It is not to engage, in a way that Husserl's account tries to do, with what I am calling 'the micro-structure' of 'going on'.

The redrawing of the metaphysics of time I am seeking to develop here, in common with Husserl, engages with events at the level of micro-structure precisely because it treats the 'taking place' or 'going on' of an event as not a mere externality but as *constitutive* of the event. This must surely accord with the commonsense view of what an event is. To think of an event is to think of something that *has* happened, or that *is* happening, or that *will* happen, or that *might* happen. Past events were once actually 'going on', and future ones (or merely *possible* future events) either will be or may be 'going on' - and the 'going on' will involve changes of tense profile on a phase-by-phase basis. Husserl's insight, arrived at by interrogating living experience, was that an event or event-phase D - say, a note within the musical phrase ABCDEFG - *points from within itself* to antecedents such as ABC, and to there being something or other 'forthcoming' (in this case EFG). D *implies* a past (ABC), and something or other (EFG, or perhaps silence) in the future. The way we speak using the continuous present form surely bears this out - for example, the statements 'it is raining', or 'D is sounding',.actually *entail* that the events described have, in each case, begun and not yet ceased, hence these statements and their like have *some* purchase on the immediate past and the immediate future.

I turn next, as promised at the end of the last section, to the significance of the additional lines in Husserl's diagram (figure 2) and of the suffixes. Below, in figure 4, I redraw and modify Husserl's diagram, both to incorporate suffixes for the present and for protentions, and in a way that invites comparison with figure 1 (for reasons that will become clear in what follows). Again, positive suffixes represent past tense profiles or retentions, zero suffixes what we designate 'present', and negative suffixes represent future tense profiles or protentions.

$$A_{-1}B_{-2}C_{-3}D_{-4}E_{-5}F_{-6}G_{-7} \quad \uparrow \text{ Past}$$
$$A_0B_{-1}C_{-2}D_{-3}E_{-4}F_{-5}G_{-6}$$
$$A_1B_0C_{-1}D_{-2}E_{-3}F_{-4}G_{-5}$$
$$A_2B_1C_0D_{-1}E_{-2}F_{-3}G_{-4}$$
$$A_3B_2C_1D_0E_{-1}F_{-2}G_{-3}$$
$$A_4B_3C_2D_1E_0F_{-1}G_{-2}$$
$$A_5B_4C_3D_2E_1F_0G_{-1}$$
$$A_6B_5C_4D_3E_2F_1G_0$$
$$A_7B_6C_5D_4E_3F_2G_1 \quad \text{Future} \downarrow$$

Figure 4

The sequence ABCDEFG is, let us assume, a melody. Line 1 shows the melody as wholly future, and line 9 (the last) as elapsed, just-past. There is a clear correspondence with Husserl's diagram (figure 2) if we ignore the 'future' part on the right. The vertical column $A_0B_0C_0D_0$...corresponds to Husserl's horizontal line ABC..., the diagonals $A_0A_1A_2A_3$...etc. to Husserl's vertical lines, and the horizontal lines (i.e. lines 1 to 9) to the diagonals in Husserl's diagram. The second of these horizontal lines represents the first note sounding, and the eighth the last note.

We start with the understanding that A, B C etc. each has some duration. Hence A_0 refers to A as 'present' in the sense that it has begun and not yet ended. (The same considerations will therefore apply to B_0, C_0 etc.) Lines 2 to 8 inclusive represent the melody ABCDEFG as present, hence the whole could be written $(ABCDEFG)_0$, the event as a whole with the suffix 'zero'. We might want to represent it in that way if we were concerned with a *larger* temporal whole - a whole piece of music of which the sequence ABCDEFG were merely a phrase, let us say. On the other hand what is represented in figure 4 as A_0 (or B_0 or C_0) *could* have a diagram like figure 4 devoted to it if what concerned us were the constituent phases of A (say $\alpha\beta\gamma\delta\varepsilon$). And, of course, since α (or β or γ) has temporal extent, a diagram of the same kind could be drawn for it too. And so on indefinitely, with the instantaneous 'now' as an ideal limit.

Following on from this, it must be clear that there is just *no such thing* as a pure presentness of the kind that St. Augustine sought but failed to find, or that the specious present theorists claimed to have found after a fashion. This is where comparing figure 4 with figure 1 can be illuminating. For what figure 1 depicts is, precisely, the impossibility of 'presentness' in the sense required - the unsustainability of what I called, in *RT*, a 'walled garden of unadulterated presentness'.[13] Such a conception is, for the reasons given earlier, logically incapable of realisation, a metaphysical chimera. In figure 4, by contrast, there is no region or quasi-region called 'the present' - there is no 'quality of presentness' which could define such a region.

What the diagonal lines in figure 2 (and the horizontal lines one to nine in figure 4) represent, then, is the successive modification which constitutes that micro-structure which (I am claiming) just *is* the event 'taking place' or going on'. While B (as B_0) is sounding, it is joined to a retention of A (as A_1) and to a protention of C (as C_{-1}). When B has given way to C, we then have B_1 as just-past, as the immediate past of C_0. This transformation involves B_1 sinking 'into the depths of the past' (as Husserl puts it)[14] with its own retained A_1 now transformed into A_2, the retention of a retention. B thus giving way to C also involves the protentional element radiating from B being transformed - realised, partially realised, or unrealised - by the arrival of C as C_0.

But what can we say about this protentional element? In addressing the issue of the future, we need to get away from secondary phenomena such as prediction, expectation etc. which presuppose an understanding of futurity, in the same way as we had to delve below representation and memory to get at a more basic understanding of pastness. The clue in the case of 'retention' and the past was to embrace the past in its positive rather than its privative aspects, to see it as a beginning, a foundation, and not a mere absence. Again, the past in its perpetual *passing* is something so inescapably tied to what we think of as 'the present' as to make the present inconceivable without it. But then the very continuity of the present that enables us to say 'it is raining', and to know that it has begun to rain and not stopped, the continuity that bespeaks the past in its perpetual passing, necessarily entails the future in its perpetual impending. A conception of the future is not derived by some abstractive process from expectation, intention, prediction etc. - rather, these are all built on the foundation of a prior understanding of the future. To say these things, whether about the past or the future, is but to draw out in detail, beyond its high level of abstraction, what is entailed by Kant's statement that 'time is not an empirical concept that has been derived from any experience'.[15] This understanding of the future can now inform our understanding of the micro-structure of an event's 'going on'. I shall return to this presently. Before I do so, I need briefly to make some methodological observations, concerned for the most part with 'the transcendental turn' in philosophy.

We need to have in mind certain consequences which follow once we own to 'the transcendental turn', whether we choose to follow Kant himself, or Kant-inspired analytical philosophers, or find inspiration in the significantly post-Kantian approaches - diverse though they are - of phenomenology. Whichever of the variety of options open to us we take here, the central Kantian insight remains - by investigating how things *appear to us*, to conscious subjects of experience, as opposed to how they might be in themselves, we are *ipso facto* investigating the things themselves in the only way we can. If we add to this the post-Kantian insight that the noumenon as Kant conceived it - i.e. as the notion of their being a *modus essendi* of what ultimately is, to which   no *modus cognoscendi* could *ever*, even in principle, correspond - is an   empty metaphysical abstraction, then 'the things themselves' are revealed as nothing other than 'things as they appear to us' (if this is not too narrowly circular).

Those who have been at all touched by Kant, both those avowing no other allegiances than to pursuing an analytical approach to philosophy (examples here would include Strawson and Putnam)[16] and those committed to some versions of phenomenology, would all agree that we neglect this central insight at our peril. Tenseless theorists clearly do so. They can so easily appear to be sitting in their dressing-gowns by their firesides with Descartes, and, like Descartes, whilst engaged in radical questioning in one sphere, at the same time betraying a blindness in relation to certain notions which remain unexamined. Just as Descartes took on board without question certain Aristotelian and Scholastic conceptions of causation, so too tenseless theorists take up certain aspects of the language of time as if they have fallen from the heavens. They march on boldly, as if theirs is the privilege of seeing reality face to face, whereas the rest of us, trapped in our 'tensed beliefs', see it in a glass darkly. They think they have gained access to 'time as it is' free from the limitations of the perspective of subjects of experience. They fail to see that they are in thrall to a minor cul-de-sac on the vast map of our temporal language, a map owing its overall structure to the anonymous community of language users.

The import of these general considerations is that what I am calling 'the micro-structure of going on' should give us not only the structure of an event and of the passage of time as these occur *within* consciousness, but the structure of events and of time as *objects for* consciousness in a public, intersubjective world. When Husserl writes of the chain of retentions as trailing off, fading into obscurity, and so on, he could be taken as offering us no more than a first person psychological report. Whilst he does offer a convincing account of the experience of the 'going on' of an event, he does much more. If we construe this account as not merely an account of the *experience* of the 'going on' of an event, but also as an account of the micro-structure of the 'going on' *itself*, we are then free to conceive the retentional chain as an extension reaching back into the past indefinitely and without beginning, and the protentional element as reaching out indefinitely into the future. We thus attain a conception of 'objective time'. Husserl's account of the process of experiencing a melody, or a single tone, must therefore be seen as an account of that experiencing which remains faithful to the pattern and order of the experienced object, the melody or tone, *itself*. That is where the account differs from the specious present account, where the unity of the temporal object is lost. To understand this is *ipso facto* to understand that the melody would have been the same even if there had been no one to witness it, hence that it occurs in an objective time.

The central idea that underpins this analysis of the micro-structure of the 'going on' of an event is the idea of intrinsic modification. This is what needs to be made clear if we are to succeed in providing an alternative picture of the 'going on' of events to McTaggart's parody of time. A further gloss on McTaggart's parody is that it consists in viewing events as self-enclosed individual entities, or substances, and their tense determinations as merely external relations. Viewing events as analogous to substances is considered by some philosophers to be the only way forward, Donald Davidson being the obvious example here. Whilst it should be clear from the course of this essay so far that this is a view I am committed to rejecting, I feel it necessary to say something about it.

Davidson's central thesis, stated in the most general terms, is that events exist, that we have to 'admit' events into our ontology - in itself an uncontroversial thesis, if anything ever is in philosophy, problematic only in the form the thesis takes for Davidson. Consider, for example, Essay 9 from his *Essays on Actions and Events*. Here, as elsewhere, his starting-point is the two ways we have of describing or reporting events. We can say either 'The death of Monteverdi was unexpected', or 'Monteverdi died unexpectedly'. The former 'names' an event, whereas the latter does not. The former 'admits events into our ontology', whereas the latter does not. To say the latter, according to Davidson, is to say that 'events as particulars may not, after all, be basic to our understanding of the world'.[17] In Essay 6, he says 'we would normally suppose that "Shem kicked Shaun" consisted in two names and a two-place predicate. I suggest, though, that we think of "kicked" as a three-place predicate' - the upshot being that for Davidson 'Shem kicked Shaun' is to be analysed as 'there is an x such that x is a kicking of Shaun by Shem', so that we have a third term or *relatum*, the event.[18] This bespeaks an understanding of the task or project of 'ontology', such that whatever 'exists' must somehow be translatable into noun-like form, and must be conceived as a thing or quasi-thing. The task of 'ontology' is then to find out what can and what cannot be expressed in this noun-like form. Whatever can is then deemed to have earned an ontological diploma, and is forthwith admitted into the sphere of 'existence'.

This approach to 'ontology' has always seemed to me to be in general very shallow and heavy-handed, and to be particularly unfortunate in this case, where precisely what we need to do is highlight the *difference* between events and things, rather than try to find ever more ingenious ways of conflating them. If events are ceremoniously admitted into 'our ontology' in this way, the question of *what* they are gets overshadowed by the ontological revelation *that* they are - and then *what* they are is deemed also to have been settled, or is deemed to be less important, when in reality it has been shirked. The event's victory in gaining its ontological diploma is pyrrhic, and the ontological ceremony itself is at best specious, at worst a kind of ontological fetishism. *Why should* the refusal to read the verb in 'Shem kicked Shaun' as a disguised three-term relation be construed as 'denying that events exist' or 'failing to accept events into our ontology'? The refusal to read it as a three-term relation, I suggest, can just as easily be construed as accepting the reality of events but denying that they are *thing-like* particulars. Further, it leaves open the question *what* events are, thus making possible a more subtle, a more nuanced, approach to the ontology of things and events. It is precisely by bringing to light the nuances that *distinguish* events from things that the real work in the metaphysics of time, space and things can be carried on in a way that measures up to the task.

I am suggesting that investigations into the micro-structure of the 'going on' of an event provide the way in to this more nuanced approach. In what follows, I take the earlier discussion further in a way that incorporates the primitive understanding of the future that is presupposed by such phenomena as expectation, prediction and intention, and that also develops further some ideas I first expressed in *RT*.

Let us begin with a comparison of figure 4 and certain aspects of figure 1 - in effect, figure 1 without the vertical lines, which would make figure 1 like figure 4 but without the suffixes. If we compare the trajectory of the notes BCD in the two diagrams, we see at once that BCD represented in figure 1 is unequivocally and in all respects *identical* throughout its changes of tense determination, which are displayed there in the successive horizontal lines (from the first to the eleventh). In figure 4, by contrast, we never have BCD *simpliciter*, but instead $B_0C_{-1}D_{-2}$ followed by $B_1C_0D_{-1}$ then by $B_2C_1D_0$ and then by $B_3C_2D_1$, and so on. What *differentiates* figure 4 from figure 1 is that in figure 4 there is a way in which D_{-2}, D_{-1}, D_0, D_1, D_2 etc. are different, and a way in which they are the same: they are different because they are *different tense profiles*, yet they are the same in being tense profiles of *the same event*. As Husserl puts it, 'the sound itself is the same, but "in the way that" it appears, the sound is continually different'.[19]

Let us look at what changes between , say, $B_1C_0D_{-1}$ and $B_2C_1D_0$. The note D is first prefigured as D_{-1} from the standpoint of the currently sounding note C (as C_0) - that is, it is necessarily the case that C_0 will be followed by *something*, even if it is only an empty interval, a silence. It may already be determined from the standpoint of C_0 what note D is, but it does not have to be. If the musicians are playing from a score, or from memory of a score, it clearly is determined; but if they are jazz musicians, it need not be.[20] Then, whichever of these is the case, D_{-1} is fulfilled as D_0. What are the important features of this change? First, a possibility opened up, or at least not closed, by C, is now *finally* closed - the chips are down, and the note improvised by the jazz trumpeter (D_0) is, for better or worse, what is actually sounding. With this change from D_{-1} to D_0, there are also the changes from C_0 to C_1 and B_1 to B_2.

For convenience of reference, and in order to illustrate a number of further points, I reproduce the relevant parts of figures 1 and 4 as figures 5a and 5b below.

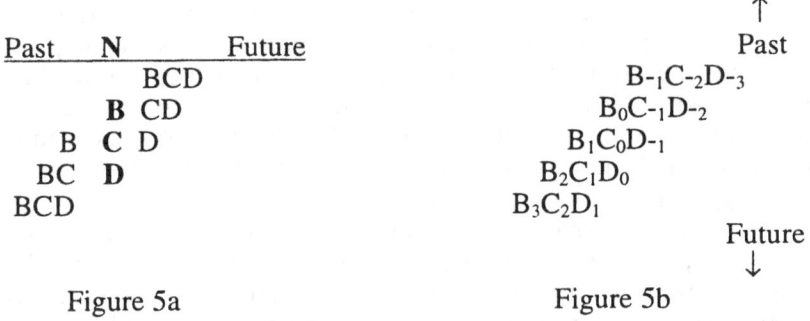

Figure 5a   Figure 5b

The first point concerns how past and future are represented differently in the two figures. Figure 5a creates a gulf between past-present-future as an eternal structure, a form, and the particular *contents*, the events, that supposedly embody that form. The notion of past-present-future as an eternal structure or form calls for explanation. Briefly, when we use the terms 'past', 'present' and 'future' we do so in two ways. First, we use terms like 'now' or 'the present' to refer to *particular* events or situations, and correlatively we speak of past and future events and situations. On this way of speaking, what *is now* past, was *formerly* present. Second, we speak of 'the present' as a generality, and in this sense 'it is *always* now, *always* the present' - and, correlatively, we have 'the always future', the future that never comes, and 'the always past'. In the former use, we can speak of the present as *nunc fluens*. In the latter use, with past-present-future as an eternal structure or form, we can speak of the present as *nunc stans*. It is, of course, essential that we preserve both these senses. The present is inescapably both *nunc stans* and *nunc fluens*. This has important consequences, as will become clear in the next section of this chapter.

Clearly figure 5a attempts to prioritise the *nunc stans* over the *nunc fluens*. This is confirmed by the basic tenet of the specious present theory, namely the idea of a fixed attentive span which the subject carries along from moment to moment. As we saw in section 2, this generates problems, the main one being that the supposed 'unity' of the attentive act is bought at the expense of the unity of the temporal object. The 'present' region of the figure, designated by the column N, is filled with successively 'present' contents - A *then* B then C - as we move down the column. All else in the diagram is either 'not yet' or 'no longer' - and as such it is totally severed from the present. The only way B, as present, can function as a locus for preserving anything of the now elapsed A is by means of some smuggled in *surrogate* of A such as a 'memory image'. Indeed, this element of surrogacy infects even what is *within* what is designated 'present'; for strictly speaking, there can be only instantaneous events in the N column. This is why the only way we can *appear* to give it content is by means of the specious present theory or its like - but then we have a 'content' to the present that consists entirely of imported surrogates (images that persist 'in the mind' etc.). This approach is, as we have seen, a metaphysical quest doomed at the outset. In seeking for its 'walled garden of unadulterated presentness', it can only *de-temporalise* time - and therein lies its impossibility.

With figure 5b, the situation is markedly different - where before we had severance, with a present specious in every sense poised between the two abysses of the past and the future, here *connection* is centre stage. If we take the transition from $B_0C_{-1}D_{-2}$ to $B_1C_0D_{-1}$ in the context of sections 3 and 4 of this chapter, we see not the *severance* of an island of presentness from two abysses of non-being, but instead *continually changing ways in which* past and future are taken up into the present. In this way, figure 5b gives us something of the structure, and the unity, of the 'going on' of the sequence BCD. In so far as it disallows the writing of B,C or D *simpliciter*, in so far as D is never anything other than what is displayed in its successive tense profiles D$_{-1}$, D$_0$, D$_1$ etc., it illustrates the element of *unsaturatedness* that characterises the internal structure of the 'going on' of an event or sequence of events.

The term 'unsaturatedness' is new in this context and calls for explanation.   I shall invoke two tendencies in Kant's first *Critique*, by way of comparison and contrast, to explain what I mean.   The first is Kant's apparently firm and immovable conviction that we cannot but view temporal order as strictly causal - upon encountering A followed by B, we are constrained *a priori* to think that A has caused B.   But then he also says that the order in which the *spatial* profiles of a thing appear to me are determined, ultimately, by me, by my own will.   Kant gives here the example of a house, which I can view in a variety of *different* perceptual sequences - I can look at the front, the side and then the back, or I can stand in front of it and view the roof, then the front door, then an upper balcony; but the crucial point for Kant is that the house does not *exist* in this successive manner.   He contrasts this with the perception of a ship moving down a river.   In that case, we are *constrained* to perceive it first upstream and then downstream.[21]   The order of events in the time of external nature is subject to strict causation, whereas whatever is the result of human volition is not (or, more accurately, belongs to the order of the causality of the will).   It would not be too fanciful a reading of Kant to say, therefore, that through space, we can buy freedom from the inexorable order of causation in time.[22]

But if this is the *only* way we can be free, we have to buy into Kant's notion of noumenal freedom, according to which exercise of the will free from the compulsion of nature is possible only as an intervention into the strictly causal order from outside it, and so the self must exist outside that order.   If we have a different notion of the self, if we view the self as at least in part *in* the world rather than contemplating it *ab extra*, and as merely one self among others, the idea of temporal order as strict *causal* order begins to lose its plausibility.

Be that as it may, the discussion of the self belongs to the next chapter, and I do not propose here to go into the theory of causation.   All that I wish the reader to take from this very brief discussion of Kant is the point that Kant recognises, implicitly if not explicitly, the need to acknowledge that we cannot divorce totally the bare, abstract *form* of time from its possible *contents*, and take the view that 'time itself' is totally indifferent to whatever might go on in it.   The argument of the second analogy can then be seen as the embodiment of Kant's recognition of that need.

The general thesis that time has at least *some* purchase on its contents may well appeal to many philosophers, but not all of these will be wedded to Kant's claims about causation.   It is possible to extract from Kant's first and second analogies, liberally interpreted, a minimalist thesis along the following lines: the flux of time, that which is the source of change, is only possible against the background of the permanent; or, put more concisely, changes *in* time presuppose the permanence *of* time. Earlier on, in the Transcendental Aesthetic, Kant has told us that time is not a *concept* that has *instances* of itself, in the way that (say) the concept 'hard disk' has a great variety of instances of it - for the moments or periods of time are all *parts of* one and the same singularity, time, in a way that could not be said of a discursive concept like 'hard disk' and its instantiations. (There is no hard disk-like singularity such that all the hard disks in the world are *parts of* it.) This is perhaps the *ultra*-minimalist thesis: time itself is inconceivable as anything other than an inescapable unity, a singularity; and yet its constituent moments admit of *any* content.

These parts of Kant's programme exist at a high level of abstraction. My purpose in invoking them is to suggest that 'unsaturatedness' as a metaphor tells us something more about the inner structure of time, of the 'going on' of events, processes and changes - furthermore, it tells us something that should take the place that is occupied, in the Kantian scheme, by the causal principle of the second analogy. Unsaturatedness creates a space that makes *significance* possible - and in a way that the purely causal language never could, it makes possible the language of time as narrative.

I shall return to the metaphor of unsaturatedness, to significance, and to time as narrative, in the course of the section after the next. The key - or at least an important clue - to understanding this, lies in Husserl's statement, quoted earlier: 'The sound itself is the same, but "in the way that" it appears, the sound is continually different'.[23]

5 The Time that Flows, the Time that Abides

I began this chapter with a twofold statement. On the one hand, we have to accept the common human experience that events 'come to pass', 'take place' or 'go on', and that time 'passes', 'elapses' or 'flows'. On the other, we have to acknowledge that giving an adequate account of these aspects of time is possibly *the* most daunting problem that philosophy has to face. In the main body of this chapter, my chief concern has been to try to account for these aspects of time, focusing on the 'going on' of an event, series of events or process. I have argued extensively that the standard attempts to account for this in terms of 'representations' such as memory images are wrong, and that in the end the *only* possible account is to be found through accepting the reality of past and future in a way that precedes and grounds any possible representation of them. The focus of the discussion leading up to this conclusion was the need to surmount the dogma that the present is 'strictly speaking' durationless, and hence that the only events that can be truly spoken of as present are instantaneous. The attack on this dogma arose from two sources: the way we use the continuous present tense, and the way we can speak of past and future in positive rather than privative terms. To take the second of these first, it was concluded that we do have some direct, non-privative experience of (at least) the immediate past and the immediate future - 'direct' in the sense that it is not by virtue of 'representations' such as memory images that we reach them, but through the originals themselves, without mediation; and 'non-privative' in the sense that, for example, the first note of the melody I am now hearing is not experienced as something lost, as merely 'no longer', but as that on which the later parts are *grounded*, that which is the *source* of what follows later. In a similar way, as I reach out to grasp something, the immediate future is not a mere 'not yet', but the directly given *meaning* of my action, like a final cause having some efficacy in the present.

The other source of attack was the analysis of propositions in the continuous present tense such as 'Peter is writing to his mother', 'it is raining' etc. Once we see the point about the continuous present tense - viz. that if Peter *is* now writing a letter, he has begun and has not yet stopped, and so there are parts of this present activity of his that can legitimately be described by the use of past or future tense statements - we can extend the analysis to cover much longer events, e.g. those corresponding to 'X is writing an autobiography', 'Y is building a house', 'Z is writing a film script'. Accordingly figure 5b, or better, an extended version of figure 4, can be used to 'map' not simply the 'going on' of a melody of six or seven notes, but equally the 'going on' of a phase of someone's life, such as 'writing an autobiography', or 'working towards a doctorate'. There may well be all kinds of *epistemological* differences between the shorter and the longer events, but as far as the *logic* of the continuous present tense goes, there is no difference at all.

Going by this logic, an event of *any* duration can be present. Of course we pay the price that we are no longer speaking of the 'psychological' present - but then this notion of a psychological present is suspect anyway, once we reflect that, for example, the fact that I am engaged on a project of large time-scale (e.g. writing an autobiography) may weigh more with me than the fact that, at this moment, I am writing a letter or out walking. The larger project may thus serve to structure the whole of my existence in a way that a succession of 'present actions' or 'incidents', understood 'psychologically', never could.

These are important considerations, and their place in this essay properly belongs to the next chapter. But it is worth raising one point even at this stage, and that concerns the way particular actions such as writing a letter, going for a walk or talking to someone may relate to the larger project. In particular, there is the question of significance: How does the smaller event or series of events in the 'psychological' present help or hinder the larger project, the project on the greater time-scale? We can imagine a number of possible scenarios here. The larger project may just get overlooked, because I may become so totally absorbed in the *minutiae* of my life that I simply lose sight of it - like some character in a play by Chekhov who starts life with grand aspirations but ends up living the life of a *petit-bourgeois*. At the other extreme, we have Pasha Antipov, in his guise as Strelnikov, in Boris Pasternak's *Doctor Zhivago*.[24] Life is somehow not big enough for Pasha, now married to Lara, of whom he feels unworthy; and in a mixture of motivations involving among them the felt need to do some great service in order to make himself worthy of Lara, he enlists to fight in the 1914 War. After a series of fortuitous circumstances, he comes to espouse the Bolshevik revolutionary cause with the zeal of the true idealist, in a manner that at once elevates his life to universal significance, and reduces its day-to-day detail to total subservience to 'the cause'. Everything is for the cause; if ruthless actions are necessary, he carries them out in its name. Meanwhile he never returns to his life with Lara and their daughter - it is as if this former life has been elevated to a higher plane, something that will only be rightfully his once he has seen this one, the life of the cause, through to its end. When 'the cause' finally closes in on him, threatening his very existence, and when he is brought to see both that Lara loved and admired him, and that he is never going to see her again, he resolves the contradiction of his life by taking it.

Between these extremes, there is the life that illustrates a different kind of resonance - one where the particular events of one's life can be gateways to another life, as in the way Anna Karenina's series of fortuitous encounters with Count Vronsky[25] lead her to enter a deeper life, and finally to its tragic end; or again, as in the case of Zhivago, whose encounters with Lara become both the source of his poetry and the sense of his existence.

I return later to the large theme that these examples touch on. My reason for invoking them here has to do with the understanding of time that they presuppose. Let us now return to the imagined indefinite extension of figure 4, with a question in mind, namely, how does the ensuing picture do justice to the 'transitory' and the 'extensive' aspects of time? These terms that I introduced in chapter one, borrowed from C.D. Broad, are self-explanatory: we speak of what passes, what is briefly enjoyed and then vanishes, and so on; but we also speak of *periods* of time in which events are ordered in a series.[26] Augustine's question about time in *Confessions* 11 can be seen as an attempt to reconcile the transitory and extensive aspects of time. Augustine is asking, in effect, how can time be the bearer of both succession and duration, both the fleeting and the abiding?

What we have, with the imagined extension of figure 4, is a synthesis of these two aspects of time. As we travel down the horizontal lines of the figure, we move from $A_{-1}B_{-2}C_{-3}D_{-4}E_{-5}F_{-6}G_{-7}$, to $A_0B_{-1}C_{-2}D_{-3}E_{-4}F_{-5}G_{-6}$ and so on, all the way to $A_7B_6C_5D_4E_3F_2G_1$. Time represented thus is, in the terms I used extensively in *RT*, a changing future-with-present-with-past conjunction. Time thus flows, yet it abides.

There is a section in Augustine's *Confessions* (book eleven), towards the end of chapter 14, where he first poses the question: how can two of time's divisions (past and future) exist, if the past 'is no longer' (*iam non est*) and the future 'is not yet' (*nondum est*)? If the present were always present, he next observes, it would not be time but eternity (*iam non esset tempus, sed aeternitas*). But if time - present time - exists *only because* it goes into the nothingness of the past, how can we say that it exists at all, since the *cause* of its being is that it will cease to be? The last sentence of the chapter poses the question: 'Can we not rightly say that time is because it tends towards non-being?' (*Ut scilicet non vere dicamus tempus esse, nisi quia tendit non esse?*) This passage, with its privative account of past (*iam non est*) and future (*nondum est*) sets the 'existence' aspect of Augustine's discussion in stone. This passage is followed immediately by the 'extent' aspect of the discussion, in which Augustine concludes that the present is a durationless instant. Putting the two aspects together, in the end Augustine is forced, as he sees it, to conclude that the 'extension' of time lies 'in the mind' and not in time itself.

But Augustine could have taken another way out, had he seen the question 'can we not rightly say that time is because it tends towards non-being', which he posed at the end of chapter 14, in a different light - had he noticed that, in effect, time is *perpetually* heading towards non-being, the future is *perpetually* going over into the past, the knife-edge present is *perpetually* overseeing the transition from one sphere of non-being (the future) to the other (the past). The word 'perpetually' should alert us to a central aspect of the terms 'now' and 'the present'. Even if we think in Augustine's terms, i.e. of the present as durationless, as vanishing, this very instant itself is *perpetually* vanishing. But it cannot be 'perpetually vanishing' unless it is also perpetually returning. The ever vanishing present is, *ipso facto*, the ever abiding present, the present that is never extinguished - its perpetual vanishing *is* its perpetual return.

But this perpetual vanishing-return cannot make any sense by itself, i.e. unless it involves a perpetual commerce with, and differentiation from, the past and the future. The question of the 'being' of the 'no longer' and the 'not yet' thus stubbornly remains. Augustine's answer - 'therefore wherever they are and whatever they are, they are nothing unless present' (*ubicumque ergo sunt quaecumque sunt, non sunt nisi praesentia*) - finally closes off any possibility of his finding an answer that will satisfy us, his modern readers. To us, his answer is wrong whatever his intentions, for the reasons given in section 2 of this chapter. If the past and the future are to be 'preserved', it must be as past and future respectively, and *not* as present. The central point, at the risk of tiresome repetition, is that the past and the future, in their very distinctness from the present - as 'absences', if we wish to make use of that metaphor - are given a *positive* characterisation: the past is, let us say, the foundation of the present, and the future its aim or direction.

I close this section with another look at figure 4. Let us consider an indefinite extension of that figure, and as before, let us view time as a movement down the successive horizontal lines of the figure. We have what I called, a few paragraphs back, a changing future-with-present-with-past conjunction. But even this is to over-simplify the situation. For there is not, in the end, a succession of discrete horizontal lines, or rather what those lines represent. One line shades *continuously* into the next. The discrete lines are each a momentary *freezing* of time, a 'still' shot, a frame in the reel of film rather than the continuous movement it aims to capture. Similarly, when we say A_0 shades into A_1 and A_1 into A_2 etc., the very *naming* of these modifications amounts to a momentary freezing of time. This should not be surprising, for it clearly follows from the earlier observation that it is possible to devote a diagram like figure 4 to one single constituent such as A_0 , thence in turn to α_0, a part of A_0, and so on without end.

6 Unsaturatedness, Significance, and the Possibility of Narrative

Much, though not all, of the preceding discussion may seem unduly abstract. It is hard to see how this could have been otherwise, given the nature of what has been under discussion. The notions of unsaturatedness, significance and narrative featuring in the title of this section can help to redress the balance. I now take up again the theme these notions gesture towards, from where I left it at the end of section 4. I suggested at that point that unsaturatedness as a metaphor tells us something about time's inner structure, something that should rightfully be accorded the same status as Kant gave to the causal principle. This suggestion calls for further explanation.

Kant's causal principle can be seen as a response to Hume, who in turn is reacting against some barely disguised scholastic notions of causation that are to be found in Descartes and Spinoza, among others. It is illuminating to view the discussion of causation among philosophers in the two centuries that include Descartes and Hume as comprising a set of responses to the threat posed by occasionalism, i.e. the doctrine that God, being all powerful, could intervene to alter the course of nature, even the laws of nature, at any time. According to this doctrine, God could institute an 'order' of nature that was in no way intelligible, an occasionalist nightmare world in which there were no principles of order, no basis on which predictions of specific outcomes could be made. This would be a world in which each event would stand in self-enclosed isolation, having no significance for any other event. Responses to this implied threat range all the way from Descartes' claim that, since God is good, he would not wish to create the occasionalist nightmare world, to Hume's claim that there are no grounds for believing in causal necessity anyway, and that what we call 'causation' is merely habitual or customary association. Kant's response to this can be seen as an attempt to silence occasionalism and at the same time put significance back into the temporal order, and he does this in a way that gives recognition to the emerging new physics of Newton. Relations of significance in general are then governed by the 'analogies of experience', and within this, specifically *temporal* relations are governed by the second analogy, the causal principle.

My claim is that, being no longer party to these earlier debates, we need to find significance in the temporal order in a far less restrictive way. It is in response to this need that I suggested the metaphor of unsaturatedness, and that unsaturatedness makes significance between events possible. To see how this works out in terms of the understanding of time I have sought to develop in this chapter, we need to remind ourselves of the way an event is manifested in different tense profiles as it sinks into the past - A_0, then A_1, then A_2, and so on - and also of Husserl's statement ' The sound itself is the same, but "in the way that" it appears, the sound is continually different'. This statement clearly corresponds to the way the event retains its identity as A, while being 'continually different' in the change of suffix, reflecting the difference of tense profile. I shall next show, using the example of a musical phrase, the appropriateness of the metaphor of unsaturatedness.

Let us consider the opening phrase of Beethoven's eighth symphony. It begins on the C an octave above middle C, and then descends rapidly through A, B flat, back to C, then down to A, resting finally on F. Suppose we entered upon a rehearsal, that we did not know what they were rehearsing, and the conductor got them to play just that first note. We would hear a note of such-and-such pitch, loudness etc., played by the orchestra. What qualities - that is, what *musical* qualities - could we ascribe to it, over and above the bare description of it as a note of such-and-such pitch, loudness etc.?   Taken in isolation, surely none; it is *musically indeterminate*. It is thus clearly appropriate to speak of the first note as unsaturated. It is only when the conductor takes the orchestra through the whole phrase that the opening note takes on its strong, affirmative character, a character bestowed on it by the notes that follow. The initial unsaturatedness of the first note has thus made possible the emergence of significance, as the rest of the phrase makes its appearance. In the process, the first note progressively acquires a distinctive musical character, a new definiteness that, hitherto, it had lacked. Analogous things can be said of the whole phrase, in relation to the phrases that follow.

In *RT*, I claimed that such cases were governed by what I called 'the principle of emerging identity'.[27] What I was trying to capture in that phrase was the way the event in question progressively acquires its distinctive character in the context of significant later events. The word 'significant' is crucial here - they cannot be just *any* events. In the present example, it is *events internal to the musical phrase* that are relevant, hence significant in the required sense, and not events that are later but wholly extraneous (such as sun-spot activity, or the noise of the traffic outside). The interesting thing about such examples in general is that they bring out in a concrete way how A_1 differs from A_2 or A_3. We may say, for example, that A, now past, *was* present. But what we *cannot* say is that A_2 was present as A_2 - it was present as A_0. This is clearly exemplified by the fact that the affirmative musical quality of the first note in our example was never present prior to its progressive emergence against the new foreground of later notes. All that 'was present' in this case is the sounding of a note of such-and-such pitch, volume, harmonic composition etc., but which was, in all other respects, as yet musically indeterminate. The first note progressively acquires its distinctive musical character in the context of the notes that follow, where previously it had no musical character at all.

Not all examples of what we might call 'emerging identity' are like this. Take, for example, Wilfred Owen's poem 'The Parable of the Old Man and the Young'.[28] The poem is a narrative - it 'tells a story'. But the story it tells is the re-telling of another, a story from the Old Testament, Genesis 22: the testing of Abraham by God, who asks Abraham to prove his faith by being prepared to sacrifice his son Isaac. The Old Testament story runs thus: God speaks to Abraham, asking him to prepare to sacrifice his son at a location to which he will be guided. Abraham prepares himself, gathers the wood for the fire, and takes a knife. At the appointed place, Abraham binds his son, places him on the altar among the wood, and prepares to slay him with the knife; whereupon the angel speaks to him, 'Lay not thy hand upon the lad'. A ram caught in a thicket by its horns presents itself, and Abraham sacrifices the ram. The angel speaks a second time, praising Abraham for his steadfast faith, and conveying to him God's words '...in blessing I will bless thee, and in multiplying I will multiply thy seed as the stars of the heavens...and in thy seed shall all the nations of the earth be blessed'.[29]

The reader of Wilfred Owen's poem is presumed to know this Old Testament story; and when he reads the first line, 'So Abram rose, and clave the wood, and went', he does so with expectations based on that story. When the poem comes to Abraham binding his son, the reader's expectations are given an added nuance, as if this is more than simply the re-telling of an old story:

Then Abram bound the youth with belts and straps,

And builded parapets and trenches there.

The angel speaks as in the Old Testament story, saying to Abraham 'Lay not thy hand upon the lad'. The angel then points to the ram, caught in the thicket by its horns, and says 'Offer the Ram of Pride instead of him'. Here we are being reassured that the old story is being re-told, but this is an uneasy reassurance - the 'Ram of Pride', capitalised thus in the poem, is not of course part of the biblical story. The poem ends with the lines:

But the old man would not so, but slew his son,
And half the seed of Europe, one by one.

So the story of how Abraham's faith won God's favour and blessing for both him and his descendants gets transformed, for us the readers as we progress through the poem, into the story of how the modern descendants of Abraham (the elders who sent the young into the trenches of the 1914 War), despite being offered an alternative sacrifice (the Ram of Pride), elect to kill their sons collectively on a large scale, transforming the blessing into a curse.   Here the Old Testament story, already clearly defined, is in a complex and subtle way transformed from within into something else - it becomes finally clear at the end that what we took to be the earlier phases of the original story are to be understood as the earlier phases of another, a more sinister story.

My purpose in drawing on this poem was to provide an example of what I have called 'emerging identity' which differs in important respects from the musical example. In the musical example, the first note is at the start totally devoid of musical qualities, and these emerge as the phrase of music progresses. In the poem, the first six lines re-tell the Old Testament story of Abraham setting out to do God's bidding. Then, in the middle of the poem, the mention of parapets and trenches begins to undermine our belief that the old story is being re-told. But when the angel speaks to Abraham in the poem, we are momentarily reassured - but not completely - that the old story is being re-told after all. It is only in the last two lines that the poet's story of the evil of the 1914 War completely takes over, that we finally know that we are being told a different story, and the hints about parapets, trenches and 'the Ram of Pride' fall into place. The example illustrates an aspect of emerging identity whereby the events in question can switch dramatically between distinct sources of significance.

The title of this section contains the words 'the possibility of narrative'. My intention in using those words was not to settle the question *whether* narrative is possible, but *how* it is possible. Specifically, I am asking: What is it *about time* that makes narrative possible? On the view of time developed in this chapter, clearly the notions of unsaturatedness and significance, which are complementary aspects of any meaningful process or series of events, must play a central rôle. The second example, the poem with its narrative structure, illustrates these points well enough.

7 Concluding Remarks

There are two concerns I wish to air in this concluding section. The first has to do with the way that time poses the biggest challenge for metaphysics, and with the sense of oddness and unreality that *any* theory of time, looked at long enough, is capable of producing. The second can be focused clearly into a single question, namely: How do we distinguish what belongs to 'time itself', 'the objective temporal order', from what belongs merely to 'the experience of time?'. I shall confine myself to some general remarks on this second concern, which I deal with first. The question, easy to pose though it may be, does not as readily admit of an answer. One reason for this is that it may be hiding two different questions, a relatively easy and a more difficult one. The first is this: How do I distinguish the time of my inner states from the time of the world at large? This is the easier of the two; and as we have seen, Kant addresses it in the anologies of experience, and more recently Strawson.[30] The other, the more difficult and less tangible question, might be formulated thus: How do we distinguish between those aspects of time that presuppose consciousness, intersubjective community, culture etc. from those that do not?

There seem to be two possible responses to this question. One is to seek for some basic notion of time that covers events, if not from the standpoint of physics, at least from that of some notion of bare facts - or more accurately, bare events - prior to the layers of convention, interpretation etc. which we (so the thesis goes) 'impose' on these bare events. If the question is asked: what would be the point of such an exercise, the answer must surely lie in some lingering hope that, somehow, we can 'construct' all that we need to say about the events of our lives out of these bare events, from the ground upwards. But this is, arguably, a false hope - for it is hard to see how we could possibly achieve the required construction without presupposing the final picture, and then the exercise would be viciously circular.

The other response is to refuse the question. We start with a certain rich conception of time; one that enables us to tell stories, create music and drama, engage in the study of history. We should arguably stay with *this*, and not with some abstraction. Perhaps the time of our plans and aspirations, our achievements and failures, the stories we tell ourselves of our origins, forms the greater part of our idea of time. It is this that should therefore be the object of our study.

Let us turn now to the first of the two concerns: the fact that the metaphysics of time is difficult, and the sense of unreality that theories of time can produce. With the metaphysics of time, language seems to be stretched to its limit - we seem to be perpetually skating on the edge of any meaningful discourse. It is tempting to play safe, and try to stay within the confines of substance-attribute discourse like McTaggart, and, following him, a whole tradition of tenseless theorists. But if what I have argued at length in this chapter is true, substance-attribute discourse is too crude an instrument with which to handle time. It is as if time gets lost between the interstices of such discourse. It is for this reason that the tenseless theory will always seem counter-intuitive to the ordinary person innocent of its philosophical sophistications, even if he or she is not able to provide any arguments against it. For it to seem at all plausible to such people, they would have to put themselves into a certain mental set, namely that of the physics student for whom something like Locke's primary quality/secondary quality distinction is unreflectively given, i.e. for whom sub-atomic particles, molecules, radiations, electromagnetic waves etc. are 'real', whilst the perceived redness, the heard sound, the odour of the blossoms etc. are either 'unreal', or 'real' in a derivative sense, namely as inner qualities, private to each subject of experience, that are caused by the former ('physical') qualities. Our imaginary student is further likely to think of time exclusively in terms of what is represented by the variable '*t*' in kinematic equations, and to think of it as simply a space-like 'dimension'. He may well be impressed by the argument that just as there is no place in 'objective space' for 'here', so there is no place in 'objective time' for 'now'. To one for whom this kind of thinking is not second nature, the tenseless theory of time will have no appeal.

Having said this, accounts of time in terms of 'passage' or 'flux' can also seem bizarre and counter-intuitive, particularly if we are hostage to spatial metaphors. Wherever McTaggart gives accounts of his A series, we are liable to get a picture of events moving towards us out of a place called 'the future', into another place called 'the present', thence into their final resting-place, 'the past'. Augustine speaks *palpably* in terms of space when he says of the past and the future, '*wherever* they are and whatever they are, they must be there as present'. From all I have said so far in this chapter, the dangers of such spatialisation of time should be obvious: once we embark on this road, we fall prey to those empiricist errors uncovered in section 2.

Of course the approach to the metaphysics of time I have adumbrated in the body of this chapter is, like anything else that involves an attempt to break new ground, prey to accusations of obscurity. What I have tried to do, with some attempts to refine and sharpen the clumsy and inappropriate tools of ordinary language, is to probe a little below the surface of the well-trodden paths along which everyday discourse moves, in the hope of retrieving some insights. It may be that the task is, ultimately, doomed - that all metaphysics ends up shipwrecked by the seemingly unconquerable, ubiquitous, yet evanescent 'somewhat' which we choose to dignify with the single word 'time'. In any case, we should not pretend that it is easy. To make the attempt, even if the insights we get are few, makes it worthwhile. One needs only to think of the difficulties posed by accepting the just-past as what Husserl calls 'a mode of originary givenness', and of the delicate intellectual footwork that is necessary if we are to avoid the various misunderstandings to which this notion is prone, to see both how daunting the task is, yet how important it is to keep trying.

Notes and References

[1] McTaggart certainly speaks in the earlier (*Mind* 1908) article in these terms, when he says that 'changes must *happen to* the events' (McTaggart, 1934, p. 114, my italics). In the corresponding chapter in *The Nature of Existence*, he speaks of the terms of the A series as relations to something outside the series. But what has passed down to the inheritors of McTaggart's language, and in particular to tenseless theorists, is the notion that 'past', 'present' and 'future' name (or rather, *purport* to name) properties or qualities.

[2] I deal with Augustine and with many of the issues raised in this section more fully in *RT* chapter 2.

[3] Edmund Husserl, *The Phenomenology of Internal Time-Consciousness*, trans. James S. Churchill, The Hague, 1964, p. 44 et seq.. (This text is hereafter referred to as *PITC*).

[4] I deal more fully with the specious present theory in *RT* chapter 2.

[5] William James took it to involve the future as well. See William James, *The Principles of Psychology*, New York and London 1890, pp. 609-610.

[6] See *RT* pp. 56-61 for other criticisms of the theory.

[7] We also need to acknowledge the way the immediate future plays a part in experience. I return to this later.

[8] In what follows, I draw on and take further some ideas I developed in *RT*, especially chapters 2 and 5.

[9] There is some dispute about whether this ascription is correct. See the discussion in David Wood, *The Deconstruction of Time*, pp. 65-7. But this need not concern us here, as the point can be made without referring to Brentano, as will become clear in what follows.

[10] The term 'originary', rare as it is in contemporary English, is a term Churchill uses to convey the idea of the *sui generis* nature of the givenness of the past, the past as 'present in person' but not 'present' in the usual senses of that word. See, for example, *PITC* p. 53: 'The intuition of the past itself cannot be a symbolization; it is an originary consciousness'.

[11] See Husserl, *PITC*, p. 49, and Merleau-Ponty, *Phenomenology of Perception* (trans. Colin Smith), p. 417.

[12] See *RT*, pp. 150-157.

[13] See *RT*, p. 65.

[14] See *PITC*, for example p. 50. The metaphor recurs frequently in the text, in various forms.

[15] Immanuel Kant, *Critique of Pure Reason*, trans. Norman Kemp-Smith, London, 1961, B 46, p. 74.

[16] See P.F. Strawson, *The Bounds of Sense*, London, 1966, passim., and Hilary Putnam, *Meaning and the Moral Sciences*, London, 1978, Part Four.

[17] Donald Davidson, *Essays on Actions and Events*, Oxford, 1980, p. 182.

[18] Ibid., p. 118.

[19] See Husserl, *PITC*, p. 45.

[20] The reader is referred back to the discussion of the future in chaper one if he or she finds this way of speaking disconcerting.

[21] Kant, *Critique* (op. cit.), B235-8 (Second Analogy).

[22] When I speak of 'strict causation', 'strict causal order' etc. in this and the following paragraphs, I am reading 'A is the cause of B' as 'A is a causally sufficient condition of B'.

[23] Husserl, *PITC*, p. 45.

[24] Boris Pasternak, *Doctor Zhivago*, trans. Max Hayward and Manya Harari, London, 1958.

[25] Leo Tolstoy, *Anna Karenin*, trans. Rosemary Edmonds, Harmondsworth, 1972.

[26] See p. 3 and the footnote reference.

[27] See *RT* chapter five.

[28] See C. Day Lewis (ed.), *The Collected Poems of Wilfred Owen*, London, 1963, p. 42.

[29] This and the other quotations from Genesis 22 are from *The Holy Bible (Revised Version)*, London, 1948, p. 23.

[30] See P.F. Strawson, *Individuals*, London 1959, and *The Bounds of Sense*, London, 1966.

3 Redrawing the Self

1 The Self: an Old Agenda in Outline

The following two propositions about the self are false: (a) that the self is distinct from, wholly above, the series of its experiences, and (b) that the self is nothing other than its experiences. Both are false, though each expresses part of the truth. I am not *simply* the sum of my experiences, but neither am I wholly *detached* from them. This amounts to a highly complex requirement for a theory of the self, and there does not seem to be, on the face of it, any way to accommodate it. The philosophical agenda informing the two propositions was set long ago, and is exemplified by Hume's discussion of personal identity in the *Treatise*, his own doubts about that discussion in the Appendix, and Kant's response to Hume in the *Critique of Pure Reason*. The story is a complex one, involving at least Descartes in addition - Hume responds to Descartes, and Kant responds to both Descartes and Hume - but I do not propose to go into the full story. My interest in the story lies particularly in the way in which Kant tries to go beyond these two predecessors.

Hume tries to get by with a conception of the self as a series or sum of experiences, but in the Appendix to the *Treatise* he has second thoughts: perceptions are 'distinct existences', and no connection between such distinct existences is perceivable. There is no place for the notion of substance, which is disallowed by the doctrine of impressions and ideas. The self cannot therefore be a 'something' standing over the stream of experiences, and neither can it be an element in that stream. In short, the ownership function is *required* by the series of experiences - yet there does not seem to be any way in which this function can be exercised.

At this point Kant comes on the scene, with a domesticated category of substance whose scope is restricted to providing a framework for what appears to us in space and time, together with a conception (if one can call it that) of the self as a necessary presupposition of the possibility of experience, of which all we can say is *that* it is, but not *what* it is - the self as *noumenon*, which we can never 'know' as an object of experience,[1] although we have a more intimate relation to it than the term 'know' could ever accomplish for us, in that we are in our moral experience, in our freedom, identical with it as will. But such a self is not in time. The series of events in the world at large, and that series which constitutes the succession of my inner states, are objects given *to* the self, and the self is therefore distinct from them and outside the temporal process.

The mistake of Descartes, viewed from the standpoint of, among others, the later Wittgenstein, or that of philosophers in the European phenomenological tradition, consists in taking the reflective standpoint - the standpoint typified by Descartes, seated in his dressing gown by the fire, engaged in the reflection which produced the *Meditations* - as somehow constituting the primary mode in which one possesses the sense of oneself, of one's identity as this person. This is then made by Descartes, in the *Meditations* and elsewhere, into a vindication of the idea of the mind (oneself in one's essential nature) as *substance*. And it is mainly this conclusion from the Cartesian starting-point which philosophers from Hume, Kant and onwards have been moved to deny. But whilst Descartes' thinking is still open to the accusation that it prioritises the reflective standpoint and thereby paints a false picture of the self or the person as a whole, there is nonetheless something of value that remains after it has been purged of all talk of 'substance'.

This valuable element lies in the way in which the *cogito* is never possible without being embedded, so to speak, in a *cogitatio*. My consciousness, once reduced to the *cogito* at the start of the second of the *Meditations*, is then explained as a series of *cogitationes* - 'thinking-about-*x*', 'perceiving-*y*' - which have no explicit reference to a real *x* or *y* lying beyond the mind. There is the notion of perception that emerges a few pages later, according to which the piece of wax is perceived as an essence, an intelligible object, and hence not by the senses but by being understood through a purely mental act (*inspectio*). And here, of course, the principal error of Descartes is further reinforced. That this is a peculiarly *intellectualised* account of sensory perception is clear if one sets it against, say, Merleau-Ponty's, where Merleau-Ponty tries to steer between the two erroneous ways of empiricism and what he calls 'intellectualism' (a term he uses in place of 'rationalism' to refer to Descartes' position and its descendants).[2] But be that as it may, the central insight lying in Descartes' *cogitationes* is the inescapably meaning-oriented nature of every instance of *cogito*. To think is to think *something or other*. There is here at least the suggestion of a notion of intentionality. But what is more important is the way in which the Cartesian *cogitationes* point back to the *cogito* to which they are essentially tied - these particular *cogitationes* could not refer back to anything else, and in that sense they are logically non-transferable. Descartes' error lay in trying to account for this essential connection in terms of substance, for which he was criticised by Hume. In responding to Hume, Kant reinstates the priority of the 'I think', but places the 'I' beyond the sphere where talk of 'substance' is possible, beyond the sphere where nouns can refer, in the noumenal realm. The transcendental ego, the noumenal self, finally prioritises and enthrones the thinking subject by putting it outside experience, outside time. We thus seem condemned to race hither and thither within the old agenda of the self, with no way of understanding the self in its pre-reflective being.

Could we perhaps read Kant differently?

2 Aporetics of the Noumenal Self

'It must be possible', says Kant, 'for the "I think" to accompany all my representations'.[3] Perhaps the crucial word here is 'possible'. This could be taken as saying two closely related things. The first is that the enterprise on which Kant is engaged is one which articulates a certain standpoint, viz. the 'transcendental' standpoint from which it is possible to uncover and discern the presuppositions of the possibility of experience in general. There is no such thing as the transcendental *subject*, only the transcendental *standpoint*. The word 'possible' in Kant's above-quoted statement serves to underline this fact, telling us further that the kind of philosophical reflection which makes us aware of our knowledge is something open to all. Again, to pursue this point further, Kant may be saying here that self-reflection is not something we are permanently and inevitably locked into, as a certain reading of Descartes might suggest, but something we can engage in by choice.

The other thing that Kant's statement could be taken as saying, closely related to this, is that Kant recognises that the origin of myself, the existing person that I am, has a pre-reflective reality rooted in everyday life, in 'lived experience'. These are not, of course, Kant's terms. But if it is only *possible* that the 'I think' should accompany all my representations, it clearly follows that it is not a necessary feature of my leading a distinctively human life that I should occupy a reflective stance for all, or even most, of my time. Kant may therefore be taken as acknowledging the priority of pre-reflective life.

Kant *could* be taken as saying this. But there is another competing claim - or so it appears. There is no thinking *substance*, no substantial self in *that* sense, and hence it is easy to take the road which says there is only the transcendental *standoint*. But this is to reckon without the phenomenon-noumenon distinction. What is given as the only possible object of knowledge from the transcendental standpoint is the *phenomenon*. The noumenon, by contrast, is unknown and unknowable. From the transcendental standpoint, the standpoint from which the basic epistemological questions are settled, the noumenon serves only as a *limiting* notion, circumscribing what we can know by specifying an outer perimeter we can never go beyond. The noumenon has a more positive sense for us in our moral experience, where it is made clear by Kant that, in so far as we can ever be said to act freely, autonomously, morally, as the 'good will' and in accordance with the categorical imperative, we *are* the noumenon, despite its unknowability.

We should perhaps remind ourselves that what concerns us here, what prompted these reflections, is the priority or otherwise of pre-reflective life. Where would acknowledging the noumenal world of freedom leave us in relation to that concern? In order to approach this question, we need to open up, to re-think, the noumenon. I now offer some thoughts on this.

There cannot be two *things* here, phenomenon on the one hand and noumenon on the other. For then we would, arguably at least, have to treat the latter as cause of the former, and 'cause' is a relation that relates what appear as phenomena - it is a category, and categories are inter-phenomenal. Phenomenon and noumenon cannot therefore be separate things; they are identical. There is, if this is not too narrowly circular, *what is*. 'What is' has a face it shows to us, as appearance, and a 'face' (for want of a better term) which, from the standpoint of knowledge, is necessarily hidden from us. The phenomenon is, in short, the appearance of the noumenon. Or again, to give it a slightly different nuance, the phenomenon is the noumenon *appearing*. But how are we to understand this? How are we to understand *the way* in which something that is essentially unknowable shows itself as knowable? What is the appearing? Why *should* it be a revelation of the noumenon rather than some kind of deception or masquerade? Isn't it just groundless *optimism* to assume the former?

An optimistic understanding of the phenomenon-noumenon distinction in the sphere of 'theoretical reason', and hence as a guide to the doing of (say) theoretical physics, would be this: the phenomenon is an 'appearance' in the sense of being a partial revelation of the noumenon, and this partial revelation is filled out by that combination of direct and indirect inquiry by which physicists pursue their subject. The world as it appears in space and time gives us some clue, in other words, as to how things really *are*, hence theory in the natural sciences need no longer be condemned to (say) an operationist construal. The phenomenon, as some kind of partial disclosure of the noumenon, can somehow give us faith that this is so, and that theoretical physics, for example, is not just a waste of time.

But what might weigh against this understanding of the distinction is the thought that it is too close to Locke's distinction between primary and secondary qualities - and reading the phenomenon-noumenon distinction in this way does not work, because Locke's distinction is already encompassed by the distinction between the categories which structure Kant's phenomenon. The categories of substance, cause and reciprocity, together with those of quantity, accommodate all that Locke would want to say about primary qualities, and the categories of quality cater similarly for Locke's secondary qualities. The noumenon must therefore lie beyond anything that falls within the scope of these categories. It lies beyond any distinction between cause and caused, between substance and accident, between agent and patient, and beyond considerations of intensive magnitude. It is similarly beyond the scope of extensive considerations, beyond the scope of 'one' as opposed to 'many', and beyond any distinction between the actual, the possible and the necessary. No categories can be deemed to apply to it. The noumenon is, as regards all theoretical activity, essentially unknowable, *wholly* other. This more austere, and arguably more sustainable interpretation of Kant's distinction, purged of all Lockean elements, leaves us at once with the following transcendental idealist conclusion: our theoretical sciences may go as far as we care to take them, we may arrive at ever more sophisticated models of the subatomic realm, ever more sophisticated cosmological theories of 'the nature and origins of the universe', ever more sophisticated understandings of genetics and of cell biology - *but we shall not thereby advance a single step towards unerstanding the noumenon.* Reality *as it is*, as opposed to reality *as it appears to us*, must necessarily remain a permanently closed book.

It is time now to remind ourselves of the question we asked a few paragraphs back, namely, where does acknowledging the noumenal realm of freedom leave us in relation to the priority or otherwise of pre-reflective life? In order to pursue this question, we need to look at the noumenon more closely from the standpoint of our moral experience.

The moral phenomenon, as opposed to the noumenon, must include within it introspectible moods and thoughts, deliberations, calculations as to what would be most 'prudent', the emergence into awareness of what I want, including all my projects, my plans, my aspirations. Within all this, I may become aware of choice situations of a distinctive kind, viz. 'ethical' or 'moral' situations where I ask myself questions like 'this may well be advantageous to me in many respects, but is it *right*?'. In exposing myself to such questions, I have the opportunity to act 'freely'and 'autonomously', as Kant understands these terms. Unless I do so, all my actions are unfree, determined - whether by other people and their expectations, or by physical, psychological or other causes which it is not in my power to influence. The existence and hence the exercise of freedom and autonomy is, for Kant, made possible because myself as it *is*, and not merely myself as it *appears* to me, exists as noumenon and unconditioned. Kant uses the term 'the intelligible realm' to distinguish the self as noumenon from the self as phenomenon, the latter lying in what he calls 'the visible realm'. The noumenal self as source of freedom and autonomy is none other than the will. Again we should remind ourselves that phenomenon and noumenon are not two different *things*. There cannot be, therefore, two distinct 'selves', a phenomenal or empirical self on the one hand and a wholly separate 'noumenal self' on the other. The two must be identical, with the former the appearance of the latter. In other words there is but one self, which is accessible from two standpoints - the conditioned standpoint of the empirical, the visible, the phenomenal, and the unconditioned, noumenal standpoint of 'the intelligible world'.[4]

But how are we to understand this? In the *Groundwork*, Kant introduces the notion of duty in the following way: he says that the concept of duty '...includes that of a good will, exposed, however, to certain subjective limitations and obstacles'.[5] These limitations and obstacles Kant puts together under the heading of 'inclinations', which, in so far as they determine our behaviour, determine us to act heteronomously and not autonomously - thus in so far as I allow the course of nature, the course of 'appearance', to determine my will, I am heteronomous. However, if I choose not to allow my will to be so determined, I am acting autonomously. In the next sentence, Kant says of the aforementioned limitations and obstacles, 'These, so far from hiding a good will or disguising it, rather bring it out by contrast and make it shine forth more brightly'.[6] The picture of the phenomenon-noumenon distinction this suggests is of the phenomenon as the external arena of the noumenon - a sphere where it expresses its nature, but in an inevitably limited and circumscribed way, and perhaps also where it undergoes a test, that of being subjected to the voices of temptation of our everyday existence.

But the phenomenon-noumenon distinction is aporetic here too. If the phenomenon is merely the *appearance* of the noumenon, how can it *make sense* to speak of bringing the 'obstacles and limitations' of one's empirical self under the command of 'the good will' lying in the noumenal self? How can the world *as it is in itself*, the unconditioned or intelligible world, operate to constrain the world *as it appears to us*? Again, if the phenomenon is merely the *appearance* of the noumenon, what is the point of morality, of moral exertion, at all? The good will abides in the noumenal sphere anyway, and has no need of the phenomenal world of appearance. Above all, it has no need to *constrain* the world of the phenomenon, no need to manifest or show itself forth in that sphere in any way at all - for what would be the point, given that it enjoys a superior status as noumenon?

3 Kant and Sartre

My first reading of Kant's *Groundwork of the Metaphysic of* Morals as an undergraduate student coincided with my first reading of another text: Sartre's *Existentialism and Humanism*. I was puzzled at the time that, whilst these texts seemed to me to be very close to each other and saying substantially the same things, neither my peers nor my mentors followed me in that judgement. I begin this section with an attempt to recapture, and where possible explore and take further, those youthful thoughts and associations. My reasons will become clear in what follows.

On the face of it, the two authors seem far apart. In what follows, I shall begin with Kant, and in particular with those aspects of Kant's *Groundwork* which are most at variance with Sartre. When I turn to Sartre, I shall again begin with what most strikingly differentiates him from Kant. Having completed these initial clarifications, I shall then pass on to where the two significantly meet.

Kant enjoins us to 'act only on that maxim which you can at the same time will that it should become a universal law', and again, 'act as if the maxim of your action were to become through your will a universal law of nature'. What makes such a Categorical Imperative possible, according to Kant, is the existence of a will which is free, autonomous, hence able to sustain actions which are not driven by one's inclinations and desires, because they typically go against the latter. This free will is, in turn, explained in terms of the unconditional, non-categorial being of the noumenon: the invisible, non-sensible 'intelligible world', of which we can say only *that* it is, not *what* it is. The nature of myself understood in this way, not as I appear to myself or others but as I am, is unknowable. Yet for Kant, the term 'noumenon' names final, ultimate being. One would expect, therefore, that insofar as I act freely, morally and autonomously in Kant's sense, my own being would not merely harmonise with or reflect ultimate being, but completely coincide with it. Were it not that, for Kant, the category of substance serves to define the structure of the world not as it ultimately *is* but only as it *appears to us*, it would be tempting to draw an analogy with Spinoza and say that insofar as I truly know myself and act in accordance with my ultimate nature, I come to see myself as identical with the one substance, God or nature (*Deus sive natura*). But at most, all that Kant will allow is that practical reason must 'think itself into' the intelligible world, but without *knowing* itself as part of such a world:

> By *thinking* itself into the intelligible world practical reason does not overstep its limits in the least: it would do so only if it sought to *intuit or feel itself* into that world. The thought in question is a merely negative one with respect to the sensible world: it gives reason no laws for determining the will and is positive only in this one point, that it combines freedom as a negative characteristic with a (positive) power as well - and indeed with a causality of reason called by us 'a will' - a power so to act that the principle of our actions may accord with the essential character of a rational cause, that is, with the condition that the maxim of these actions should have the validity of a universal law. If practical reason were also to import an *object of the will* - that is, a motive of action - from the intelligible world, it would overstep its limits and pretend to an acquaintance with something of which it has no knowledge. The concept of the intelligible world is thus only *a point of view* which reason finds itself constrained to adopt outside appearances *in order to conceive itself as practical.*

Kant is here denying a central idea of what we have come to think of as Platonic tradition, namely the idea of *noêsis* as a positive intellectual intuition of the intelligible realm of Platonic forms, real essences or whatever - an idea which arguably survives into the seventeenth century as, for example, Spinoza's *scientia intuitiva*. The break with the Platonic tradition here is effected by the noumenon, of which we cannot have a positive intellectual intuition, a *noêsis* - that, at least, is the claim. Simultaneously, the self as it *is in itself*, not as it *appears*, is located in a Platonic realm which can never be a *noêton* or 'object of' *noêsis*. Kant's break with the tradition is thus, to say the least, ambiguous - we possess, indeed *are* in our innermost essence, the noumenon; yet the noumenon remains a *noêton* in what is very much a 'would be' sense, object of a *noêsis* of which we are declared to be radically incapable. Yet the noumenon remains a central pillar of the Kantian enterprise. Without it, freedom of the will, and hence autonomy, is impossible.

Nowhere is Kant more at variance with Sartre than here. For whilst Kant would agree that the noumenal self does not have a 'nature' in the sense of a set of qualities inhering in a substance, i.e. an essence which can be understood in categorial terms (or indeed known in any way at all), it is nonetheless real for him in a more fundamental way than is the knowable world in space and time. This has to be so, if the noumenon is to be taken seriously. For Sartre, on the other hand, there is no such reality - no hidden, unkowable metaphysical ground to support the self. There is, instead, a void, well expressed in *Existentialism and Humanism* in such general statements as 'existence precedes essence', 'man is what he makes of himself', etc. This brief essay was written after Sartre's radical break with Husserlian phenomenology, marked by the publication of *Transcendence of the Ego* and later *Being and Nothingness*; and the crucial notions which these general statements point towards are those of being in-itself (*être en-soi*) and being for-itself (*être pour-soi*). These notions constitute Sartre's distinctive adaptation of the notion of intentionality. Sartre claims that for Husserl, the object of the thought or other mental act is internal to the act, in a way which leaves him open to the charge of idealism. He clearly thinks that there is in Husserl an unexamined, and essentially Cartesian, notion of *cogitationes* and their interiority. Sartre, by contrast, wants to say that what constitutes intentionality is not that there are contents *in* consciousness, but that there are objects *for* consciousness. This is what forms the basis of what we might call his 'fundamental ontology' (the term is not entirely inappropriate, as Sartre owes a debt to Heidegger) understood in terms of the aforementioned distinction between being for-itself and being in-itself. Being for-itself defines itself in relation to the latter in a special way: not by difference, contrast or logical implication, but through a 'nihilation' by which it interposes a 'nothingness' between itself and being in-itself. I shall explore this below, after some preliminary remarks about intentionality.

'To seek is to seek *something* (or *someone*)', 'to believe is to believe that *p*'; and the 'something', or the 'someone', or the proposition '*p*', (a) may but need not exist or be true (I may be seeking the elixir of life or for Zeus, or I may believe that there are leprechauns), and (b) carries no referential implications (if I am seeking Zeus, it doesn't follow that I am seeking Jupiter, though the two terms refer to the same deity). This is, by and large, how intentionality is understood by analytical philosophers. To understand the rôle this notion plays in Husserl, let alone Sartre, requires some additions. It requires us to see intentionality as more than a specialised kind of linguistic device which exists within, and is subordinated to, extensional or 'objective' discourse. To begin with Husserl, we have to include the notions of the epoché, 'bracketing', suspending 'the natural standpoint' etc., the upshot of which is to widen the notion of intentionality so that it goes beyond particular acts of thinking, seeking, loving and so on, to encompass the whole sphere of our experience. All language, all experience, then becomes intentional in this wider sense. The world as a whole, including many of our practises, institutions, and no doubt what Wittgenstein would later call 'forms of life', becomes the fundamental 'intentional object' - i.e. what is there *for* the pure consciousness of the transcendental ego, as the field for phenomenological inquiry.

Where Sartre differs from Husserl is over the transcendental ego and its constituting activity. For Sartre, consciousness finds itself tied to a world from which, at the same time, it finds itself distanced. This distancing is called by Sartre 'nihilation'. Being for-itself perpetually defines itself as, precisely, *not* the particular experience which it is subject to at the moment. It is perpetually disidentified with, hence radically free in relation to, any determinate being of whatever form. It seeks to attain some definiteness, some determinate character, only to find itself thrown back on its own freedom: 'The For-itself is the being which has to be its being in the diasporatic form of Temporality'. I shall not, at this stage, go into the 'ecstatic' account of temporality which is at work here - I shall return to it later - except to note that it makes clear the way in which its freedom is inescapably thrust onto being for-itself from moment to moment.

Let us return now to Kant. For Kant, freedom is inescapable because I as I *am* in myself and not as I merely *appear* - I as noumenal self - am not determinate in terms of space, time and the categories. The self as noumenal will is, to all intents and purposes, as radically indeterminate as Sartre's being for-itself. For Sartre, man is condemned to be free, condemned to invent humanity in his every act or forebearance. In a similar vein, it could be claimed that Kant's noumenal self is condemned to invent humanity as a Kingdom of Ends. The only difference is that Kant's 'invention' of humanity is through the categorical imperative - the *negative* test of 'could this be willed consistently as a universal law for all rational beings'; and the *positive* test of 'is this consistent with, or does it promote, humanity as an end in itself'. For Sartre, in living my life in this *particular* way, in making those choices which are inescapably *mine*, I am 'inventing' humanity but without the framework of those 'laws of freedom' which consitute Kant's version of the Enlightenment project. Take these away, and what is left is (at most) 'the good will' - except that there is no 'good' to fall back on either. But subjectivity as being for-itself has, despite (or perhaps even *because of*) this poverty of being, the universalising *intent*, albeit an anguished intent.

In the end, perhaps what they share is at least the spirit, if not the substance, of the 'formula of autonomy'. I inescapably see myself as the *maker* of that 'moral law' to which I am also subject.

4 The Self and its 'Objects'

In what follows, I take up again the theme of the priority of the self in its pre-reflective being, and the relocation of the reflective standpoint which this priority entails. That we need to free ourselves from the Cartesian privileging of that standpoint and redraw the self from a more secure one is a position that has been widely canvassed, and one that commands widespread support. In this and the sections that follow, I offer some thoughts on redrawing the 'logical geography'[7] of the self, focusing particularly on character, the will, reason, and the notion of a disposition. I shall call upon a number of sources for inspiration. These will include critics of 'thin' or 'unencumbered' conceptions of the self broadly within the analytical tradition in philosophy. The ones I have in mind here are Michael Sandel and Bernard Williams. I shall also draw on the early Heidegger (mostly *Being and Time*), Hegel's *Philosophy of Right*, and certain themes in Plato and Aristotle. Heidegger's 'Preparatory Fundamental Analysis of *Dasein*' (the title of *Being and Time* Part 1 Division 1) will inform much of what follows, even if I do not refer to it directly. I draw on Hegel for his discussion of the will. The Greek themes will include the tripartite division of the psyche in Plato (mainly the *Republic*), to which I shall refer directly. The discussion of the ethical virtues in the *Nicomachean Ethics* of Aristotle will be in the background. My aim is not to undertake anything like exegesis of any of these texts. I simply draw on what seem to me to be useful insights

I begin with the need to break free from the two opposed but equally false propositions which opened section 1 of this chapter: (a) that the self is distinct from, wholly above, the stream of its experiences, and (b) that the self is nothing other than the stream of its experiences. As a further preliminary, I draw a distinction between two kinds of inner object given to the particular self. The term 'inner object' is, I hope, self-explanatory - it is intended to cover everything from one's own sensations (of which one is inevitably aware) to one's traits of character (of which one *might* become aware). The term itself, and the distinction I build on it, are not intended as 'final categories of thought'. They are provisional. They are, in fact, born of the very conception of the self which, among other things, would privilege the reflective standpoint. As will become clear, there is something highly artificial about calling inner objects of the second kind *objects* at all, and something equally artificial about calling them *inner* objects. The aim of the discussion that follows is, then, to show the inadequacy of the conception of the self presupposed by these terms, partly through exposing the inadequacy of the terms themselves.

The two kinds of inner object are distinguished as follows: (a) On the one hand, we have sensations, thoughts that just occur to us, sudden feelings such as irritation, momentary anger, and other inner episodes. Though diverse, these items do not constitute a list of *all possible* inner episodes. The list is meant only as illustrative. What they share is the property of being self-disclosing - one cannot have them without being aware of them. (b) On the other hand, we have psychic dispositions or traits of character, attitudes of various kinds (e.g. value commitments, artistic preferences, attitudes to particular people), unconscious thoughts and feelings, life projects, and much that is either unnameable or not easily named, but which might be called, following Sandel, 'constitutive attachments'. Again, these constitute a diverse but not exhaustive list. They have in common the feature that, although they are mine in the same logically non-transferable sense as are my episodic feelings, sensations etc., they need not be *present to* my consciousness. I may, for example, remain in total ignorance of a character trait of mine which has been common knowledge among my friends. To take a different case, it remains true of Paul that he loves the music of Chopin, despite the fact that he is at this moment working at his desk and not listening to it, thinking about it, playing it or singing it to himself. Again, consciously articulated life projects are, according to psychoanalytical thinking, masks for other unacknowledged agendas which are ours, but of which we remain for the most part ignorant.

Any close scrutiny of inner objects of the latter kind must bring into question a certain picture of the self. This picture constitutes the major underpinning of the Hume/Kant agenda on the self discussed earlier. The picture is, roughly, this: On the one hand, we have a succession of experiences or inner objects. In so far as they are known, they must be *objects* of experience, hence not the *subject* of experience. On the other hand, we have the subject of experience itself, which, as subject, can never be *object* and hence can never be known. Such a picture both Hume and Kant would happily endorse. Hume concludes from this, in the *Treatise*, that the idea of self is not a genuine idea, 'for from what impression cou'd it be derived' - although he later expresses, in the Appendix to the *Treatise*, his famous dissatisfaction with these conclusions. What Kant concludes, on the other hand, is that we must presuppose *that* there is a self which owns the experiences, but we cannot specify the '*what*' of that self, we cannot say anything about its nature or essence - and he shares with Hume the belief that it cannot be a substance, though for very different reasons.

Following Shoemaker, I draw this picture by means of the schema 'S→O', where S denotes the unknown (or unknowable) subject, and O whatever the subject knows. And I want to suggest that thinking seriously about the nature of inner objects of the second kind exposes the weakness of this picture.[8]

The S→O schema is what we would draw if asked to draw (say) a man looking at a tree - it is peculiarly germane to *visual* perception. Yet, as Shoemaker noted, the picture is extended to cover images, sensations, thoughts, indeed anything which may be deemed an object of 'acquaintance' in Russell's sense. Human existence is represented as in essence a self (S) which stands in this relation of acquaintance to a mental object (O) or set of mental objects (O_1, O_2, O_3 etc.). The S→O schema thus gets elevated into a metaphysics of the self which powerfully influences, and sometimes even dominates, the philosophical tradition from Descartes through to Kant and beyond.

That the picture is inadequate even for, say, a man in pain, is clear after even a moment's thought. My relation to my own pain is wrongly pictured, if it is construed in the detached way the S→O schema suggests - I am not merely *acquainted* with my pain, and hence it is wrong to think of it as an object *in the same way* as an object of sight. It is tempting to say, 'it is not an object at all'. But if it is not an object, then (on the S→O picture) it must be the subject - if it is not O, it must be S. No other possibility exists on that picture. And if we turn to inner objects of the second kind, the same thing holds: if I come to know that I have a certain trait of character, what I thus come to know is not *merely* an object at a distance from me; but neither is it *merely* the essentially unknowable knowing subject. No other possiblity exists on the S→O picture. To envisage other possibilities, we have to redraw the self.

According to the S→O schema, knowing one's own feelings, dispositions, traits of character etc. would be a simple matter of introspection, directing one's attention appropriately.   But there are difficulties in conceiving one's dispositions, character-traits, attitudes, submerged or unconscious feelings etc. in this way. Firstly, a character-trait or attitude (to take these cases) is not an introspectible object which can be said to exist at a particular moment anyway, and so is unlikely to come furnished with an infallible sign of itself for the introspective gaze. The nature of Swann's 'love' for Odette is certainly not something which he can discern by introspection - and does he *ever* know its true nature?

Secondly, the relationship of the self (or of *one*self) to one's dispositions, buried emotions, unspoken yearnings, constitutive attachments etc. is too close, too intimate, for the S→O schema to make sense. To take an example: Anna Karenina comes to realise she is in love with Vronsky. Let us then speak of 'Anna's love for Vronsky' as an inner object of the second kind What form does her realisation that she loves Vronsky take? As portrayed by Tolstoy, it is a gradual dawning. Their first meeting is at the Moscow station. Here Tolstoy conveys to us the way each is drawn to the other in an unspoken way. Next, they catch sight of each other from afar when Vronsky calls briefly on the Oblonskys - with a look of dismay mixed with embarassment on his part, and a sensation of pleasure mixed with apprehension on hers. Next, they meet at a ball, where Kitty notices the way both Anna and Vronsky come to life in each other's gaze - and so the *idea* of their love exists fully for the first time *in the eyes of another,* viz. the jealous Kitty who is herself in love with Vronsky. After leaving the ball, Anna resolves to return to St. Petersburg the next day.

In the first part of the return journey to St. Petersburg, in the midst of changing emotions that range from joy to shame, she tells herself that Vronsky is nothing to her. The train stops in the night at a station, and Anna alights for fresh air. She and Vronsky meet again. Vronsky speaks directly of his feelings for her. She rebukes him. 'Though she could remember neither her own words nor his, she felt instinctively that that brief interchange had drawn them terribly close together; and this both frightened her and made her happy'.[9] When the train arrives in St. Petersburg in the morning, and her husband Karenin comes to meet her, we perceive through her eyes a distance between them. First, she notices some absurdities in his appearance and manner. Then she becomes aware of a feeling of hypocrisy - and, even though she had not been aware of it before, that feeling nonetheless presents itself to her as both familiar and characteristic of their relationship.

I shall not follow this futher except to note that, as their love develops, it becomes less and less possible for Anna to view it as an externality, as something she can distance herself from. Discovering she loves Vronsky is not, and never was even remotely, discovering in herself some introspectible *object*, an O which she, the subject S, merely *observes*. In discovering her love, she undergoes an awakening, comes to see herself in a different light. She does not 'choose' to love Vronsky, yet in the emerging relationship to him, her will both enacts and discovers itself.

At this point it should be apparent how unequal to the burden placed on it by this example is talk of 'inner objects of the second kind'. The new reality, with all its new joys, and new despair which places both her and Vronsky beyond the possibility of 'happiness' - the dawning realisation of where she has come to in her life - is at last, for Anna, an object for reflective consciousness. But this reflective awareness merely bears witness to what has already been sought after, willed, enacted, discovered.

Whatever else it may tell us - to return to the opening theme - this example bears out the need to acknowledge the priority of the self in its pre-reflective reality. Substantially the same point can be made, and has been made, in many ways. One can speak of the centrality of character, and the priority of character over any notion of a universal rational self which is capable of willing the universal maxims of a Categorical Imperative or a General Will (as Bernard Williams does). Or one can speak (following Sandel) of a 'thick' as opposed to a 'thin' conception of the self, where 'constitutive attachments' define not merely what I will, what I do, or what I have, but what I *am*.

I explore some of these in the sections that follow. I begin, in the next section, with the notion of the will.

5 Will and the Sartrean Self

A common perception has been that Sartre's for-itself/in-itself distinction places the self perpetually at a distance from its projects, and makes 'constitutive attachment' in Sandel's sense, or 'categorical desire' in Williams's sense, inconceivable and impossible. But I am not altogether happy with this, and this doubt comes to the fore for me particularly in regard to the other. A qualification of the phrase 'placing at a distance' is needed.

It is natural to think of the other in Sartre as another being for-itself which I encounter as a 'look' - I am *seen* by the other, and this points to the conclusion that I *see* the other because I am *seen* by him, and am impelled to turn the other into an object for me in order to 'fight off' being turned into an object for him. But the relation is more intimate than that. The look of the other, as mere possibility even, constitutes the first genesis of the idea of myself as actually *inhabiting* the world. This is because, for Sartre, the 'fundamental ontology' is: being in-itself, being for-itself, and being for-others (*être-pour-autrui*). Being for the other, hence being seen by or being in the gaze of the other, is thus not a 'something' *for* consciousness, not an object, not something having the character of being in-itself, but *a fundamental ontological structure*. But, insofar as I am *seen* by the other, I cannot simultaneously *see* him or her. For to see the other is to see him as simply in the world, as an object; and to see the other in this way (and there *is* no other way) is to *abolish* the 'gaze' or the 'look'.

In short, it is right to think of self/other as an irresolvable struggle, a master/slave dialectic with no resolution or forward movement. But we need to accord sufficient weight to the peculiar *intimacy* of 'the look', the way it reaches me at my very core. How is this relevant to the issue of the self and its constitutive attachments? Briefly, as follows: perhaps we need to read Sartre, not as simply *denying* such attachments, but as offering a radically different *account* of them. What I shall be arguing in what follows is that the account does not work, for reasons that I shall try to make clear.

At the heart of that account is a certain conception of the will, a conception wider than what is common among analytical philosophers. Analytical philosophers, if they speak of the will at all, tend to be informed by Descartes rather than by (say) Spinoza. Descartes' conception of the will is preserved in Rousseau and in Kant, whereas Spinoza's is passed on through Hegel. Kant, for example, begins the *Groundwork of the Metaphysic of Morals* with the assertion that the only thing that is good without qualification is a good will. Qualities of character, he goes on to say, whilst praiseworthy and valuable in many ways, are not absolutely good in the same way as the good will - a man may have fine qualities of character such as self-discipline and courage, for example, and yet be of evil will, and hence an evil man. Only the good will constitutes the true self - all else is secondary. (In precluding knowledge of one's own character by placing it behind the veil of ignorance, Rawls too is arguably following this perception of Kant.)

With this marginalisation of character, there goes a particular conception of the will. Briefly, this conception is as follows: the will is the constraining power which transforms ethical judgements into constraints on inclination, and it is in that sense an ally of reason. Both Rousseau and Kant subscribe to a divided self - a higher or (in the case of Kant) noumenal self which is the unconditioned source of the good will, or (for Rousseau) the General Will, the self that commands and controls, and another which is commanded, controlled and subdued. Now the Cartesian inheritance should be clear: the will is conceived as an essentially blind faculty which either executes the judgements of intellect, or withholds assent. It has no intelligence of its own, but merely actualises what the parliament of the mind has chosen. This conception of the will is paradigmatically exemplified in the cycle of planning, decision, execution. What is 'willed', is willed under a description - 'I will just precisely this, and nothing else'. We can call this 'the narrow conception'. But there is another, which we can call 'the wider conception', which pervades (for example) the writings of Spinoza and Hegel.

On the narrow conception, the only will that is admissible is the will directed to action, the will to do precisely this or that. But this devalues two other human concerns - the will to be, and the will to understand. On the narrow conception, these are reduced respectively to the will to possess, and the will towards intellectual appropriation for instrumental purposes. With the wider conception, the will pervades the whole psychic life. It does not simply issue forth in imperialist/dictatorial 'acts of will'. It is not directed exclusively to predescribed ends. On this wider conception of the will, it is easier to understand such quintessentially human aspects of our lives as artistic creativity, the pursuit of truth, openness to experience and to other people, uniqueness, love and so on. Most of our deeper attachments to life are lived through this wider conception of the will: anything creative, any deep human relationship, any genuine commitment to abiding values, and anything having the character of true community.

Essentially this wider conception of the will is as much characterised by assent, acceptance, openness etc. as by trying, struggling and so on. And it should be easy to see that it can readily accommodate the unconscious elements of our experience. But from the standpoint of the narrow conception, all this must appear as an abdication of the will - not only because there are no specific 'acts' of will with specific, datable outcomes, but also because the very language of the wider conception (acceptance, openness etc.) seems to be the antithesis of will.

But this wider conception of the will is in reality more in harmony with ordinary ethical experience, which in turn calls for a more adequate understanding of the self - an understanding which abandons the rigid division between a universal, rational self and a particular, 'lower' self, and at the same time brings back character to the centre of the self. Clearly the Kantian noumenal self provides a strong conception of personal identity through time - but we pay a price. Such a 'pure' self cannot be changed by experience; the ethical sphere remains 'a perpetual ought-to-be which never is', unable to be appropriated by the particular self. And, as Hegel saw, if we try to take refuge in the opposite pole of the divided self, we simply replace the perpetual self-sacrifice of duty for duty's sake with an unending series of desire satisfactions with no unity or overall direction - a self-defeating exercise in 'pleasure for pleasure's sake', in F.H. Bradley's famous phrase.

Sartre's *Being and Nothingness*, while endorsing the wider conception of the will just outlined, at the same time contains what we have come to think of as quintessentially Sartrean elements which are at odds with it. That he endorses the wider conception of the will should be clear from the language of acceptance, from the language of a responsibility which is inescapable, from the assertion that 'we are condemned to be free', and from numerous assertions to the effect that even when not choosing between two courses of action *A* and *B*, I am nonetheless exercising my will

- I cannot, so to speak, 'take a holiday' from the will, for even when I am doing nothing, I am acquiescing in a certain situation, a *status quo*.

But once we reflect on the radical nature of Sartre's distinction between being for-itself and being in-itself, we may come to see that this distinction cannot sustain the wider conception of the will - in that it is hard to see how that wider conception is *possible* if we take Sartre's distinction seriously. My argument for this conclusion is, stated in two sentences, as follows: First, in so far as I am identified with any of my 'constitutive attachments', 'ground projects' or 'categorical desires', this must appear as a weakness for Sartre - the weakness of self deception or 'bad faith'; and second, if we are to take such notions as constitutive attachment, ground project and categorical desire at all seriously, we have to concede that what these terms point towards is something quintessentially human, and should therefore be counted as a human *strength* rather than a weakness.

To expand on these two sentences, the first tells us this: If I *could* distance myself from whatever terms like 'ground project' are intended to single out, if I *could* see them as mere externalities, and if this were the only way of avoiding the charge of self deception or 'bad faith', they would not *be* constitutive attachments in the sense required. In short, so long as these constitutive attachments constitute what I *am* rather than what I *profess* or merely *have*, I am in bad faith. But if I succeed in *avoiding* this charge, I no longer have constitutive attachments.

To explicate the second sentence requires a 'longer way', to borrow words spoken by Socrates in Plato's *Republic*.[10]

We need some notion of selfhood that makes possible, or understandable, those conditions or states of mind, or states of being, or aspects of oneself such as one's character, in which I am neither totally detached from some part of my experience, nor so totally identified with it that it leaves me totally captive, totally absorbed without remainder. This seems an impossible requirement; but we can begin to approach it through taking up again the discussion of the will. I propose to do so through the discussion of the will in the Introduction to Hegel's *Philosophy of Right*.

But first, I have some further comments on the account of freedom at the heart of Sartre's ontology of the subject. This ontology carries with it a view of freedom as inescapable, and of the subject's ultimate and final responsibility not only for its 'choices' in the narrow sense, but also for its way of *being*, and even (in certain respects) the 'world' in which this subject makes its uneasy peace and pursues its projects. This view of human freedom constitutes, for those who are sympathetic to Sartre's ontology of the subject, perhaps its most appealing aspect. *All* my doings, choosings, refrainings, acquiescences, indeed the very *style* or *mode* of my being-in-the-world, are in this sense part of my freedom, hence part of my 'will'. There is thus embedded in Sartre's account of the subjectivity of the subject, although it is not explicitly *avowed* as such, what I have called 'the wider conception of the will'. But, I claim, this wider conception of the will is at odds with, cannot be accommodated by, Sartre's ontology of the subject.

In an earlier paper, I argued that Sartre's theory of the self fails because it involves a perpetual and unresolvable tension between two radically opposed conceptions of the self and its experiences, by which first the one conception seems to be embraced, and then the other seems to be emraced, and so on in continuous oscillation.[11] The two conceptions are: (a) the self as totally *beyond* its experiences, separated from them, and (b) the self as *nothing but* the series or sum of those experiences. By the first, I am totally separate from my experiences, by the second totally merged with them. Temporality promises me, first, a kind of determinate being as I act in the world, and then leaves me radically separated from that being. But neither temporality's promise, nor its betrayal, can be sustained as a final position. I shall return to this presently. What is immediately clear as regards the will, however, is that the for-itself/in-itself ontology forces on us the narrow conception: I choose just *this* course of action, and thereby forego *that* one. My will is thus nothing but a faculty directed to predescribed ends. What then has become of the wider conception of the will which, in another breath, Sartre has sought to endorse? I turn now to Hegel, who, I claim, can offer us a way out of Sartre's seemingly unresolvable contradictions.

First, it is important to emphasise that for Hegel the will is *essentially* free. It makes no sense to give first an account of the 'nature' of the will, and then ask the question, Is the will free?, as if this were a *further* question. Hegel is working with what we have designated 'the wider conception of the will', and on this conception the terms 'will' and 'freedom' are effectively synonymous. And it is only on this wider conception that such statements as 'will and intellect are one and the same thing' (Spinoza) and 'will...is the truth of intelligence' (Hegel) can make sense.[12]

I turn now to the detail of Hegel's account, which is to be found in paragraphs 4 to 7 of the Introduction to the *Philosophy of Right*. For Hegel, any notion of the will that does not include three essential elements, which he calls its three 'moments', is incomplete and paradoxical. The first moment of the will is the universal understood as negation of everything particular, and is best understood as totally unrestricted 'freedom from', a radical disengagement with all particular acts and concerns. Much of what Hegel wants to convey here is expressed very well in Sartre's account of consciousness as being for-itself. Hegel's point about it is that while it is essential to any account of the will, by itself it is an 'abstraction' in Hegel's peculiar sense of this term: it is one-sided, it contains *part* of the picture, and by itself generates contradictions which point towards the need to take *another* notion on board. While 'freedom from' gives us an essential insight into the will, it leaves it without any content at all. Reason hence presses us forward to what Hegel designates 'the second moment of the will': to will is to will *something*, and any such 'something' must be *particular*. But this too is unsatisfactory for Hegel. To will just something *particular* is to be limited by that particularity, tied down and imprisoned by it.

Thus without any further elucidations, we are for Hegel left with a will which is at once total liberation, total 'freedom from' every attachment or choice, and at the same time total limitation, inescapably tied to *this* particular choice, action or attachment. It is a 'free' choosing, yet a choosing of just *this* and nothing else. It is the most radical unburdening of oneself, and at the same time the most final shackling of oneself. It is, in short, a standing contradiction. Sartre's account of consciousness - as a perpetually sustained and unresolvable playing out through endless time of the relation (or, perhaps better, *non*-relation) between being for-itself and being in-itself - is perhaps nothing other than a dramatisation of this standing contradiction. The will is both liberation, or escape, and at the same time a choosing or assent to what subjectivity defines itself as standing over against, and hence liberating itself or escaping from. To be a subject is, for Sartre, to be perpetually *on the way* to being in-itself, perpetually *seeking to coincide* with one's chosen aim or way of being, only to find at the heart of this embrace, in its utmost intimacy, that peculiar distancing characteristic of being for-itself. I can only remain free by having a means of escape from the particularity, the 'this', which confronts me - and I can only *exercise* that freedom, *realise* it, by seeking to *bind* myself to the 'this'. But this situation can only be maintained through a way of being which leaves these contradictory requirements perpetually at play with one another, namely the *temporal* way of being: 'The For-itself', Sartre tells us, 'is the being which has to be its being in the diasporatic form of Temporality'.[13]

Hegel claims to take us beyond this in the 'third moment' of the will, explicated in paragraph 7 of the Introduction to the *Philosophy of Right*. He speaks here of the will as 'the unity of both these moments' (viz. the first two), in what must initially appear to his readers as a re-statement of the problem: for precisely what we *want* is some way of comprehending together the different intuitions to which the first and second moments speak. Hegel goes on to qualify this, as follows: 'It is particularity reflected into itself and so brought back to universality, i.e. it is individuality. It is the *self*-determination of the ego, which means that at one and the same time the ego posits itself as its own negative, i.e. as restricted and determinate, and yet remains by itself, i.e. in its self-identity and universality'.[14] We are then told, a little futher on, that this third moment 'is the one into which the Understanding fails to advance, for it is precisely the concept which it persists in calling inconceivable'.[15]   In what way does Hegel offer us a *solution* to the problem posed by the first two moments, particularly as he goes on to tell us that 'the Understanding' - i.e. everyday, discursive, *pre*-philosophical (or pre-Hegelian) thinking - cannot even *conceive* the proposed solution?

This is difficult terrain. But some hope is offered by the Addition to paragraph 7, where Hegel gives us an example which, if we can *see* it in a certain way, may allow the arcane-seeming language of the dialectic to speak to us. The example is the feelings associated with friendship and love. My paraphrase of this example, and its force, is as follows: In friendship, and even more so in love, I restrict myself to this one particular other person - out of all possible women, let us say, I have become drawn to just *this* one. If my 'choice' is authentic, if it is a choice embodying 'true friendship' or 'true love', I will not feel it as a restriction but rather as a liberation, giving me an enhanced sense of my own selfhood: 'In this determinacy, a man should not feel himself determined; on the contrary, *since he treats the other as other*, it is there that he first arrives at the feeling of his own selfhood'.[16]

Later on, Hegel speaks again of love. In the Addition to para. 158, he says:

> Love means in general terms the consciousness of my unity with another, so that I am not in selfish isolation but win my self-consciousness only as the renunciation of my independence and through knowing myself as the unity of myself with another and of the other with me.....The first moment in love is that I do not wish to be a self-subsistent and independent person and that, if I were, then I would feel defective and incomplete. The second moment is that I find myself in another person, that I count for something in the other, while the other in turn comes to count for something in me. Love, therefore, is the most tremendous contradiction; the Understanding cannot resolve it since there is nothing more stubborn than this point of self-consciousness which is negated and which nevertheless I ought to possess as affirmative. Love is at once the propounding and the resolving of this contradiction. As the resolving of it, love is unity of an ethical type.[17]

While each of the two Additions from which I have quoted serves a different purpose, they do share the common conclusion that love is a *unity* with another, in which it can be said not merely that my individuality is not lost, but that it *finds* itself. Indeed Hegel's claim seems precisely to be that I pass from a defective and incomplete state of isolated particularity and find, in unity with another, my *true* individuality.

The point that is being made here, if we take it as primarily a point about will and self realisation, can be generalised beyond the particular example of love. The clue lies in the Remark to para. 124 in the section on Morality, where Hegel first speaks of 'the right of subjective freedom'. The basic idea is that the side of our nature that is the source of particular needs and aspirations is legitimate and ours by right. It is not something to be demoted or dismissed in the name of some abstract, 'higher' morality, not something *dictated to* by a 'General Will' or 'Categorical Imperative', but rather something to be *harmonised* with such universal requirements. Hegel introduces the right of subjective freedom in the following words:

> Amongst the primary shapes which this right assumes are love, romanticism, the quest for the eternal salvation of the individual, &c; next come moral convictions and conscience; and finally, the other forms, some of which come into prominence in what follows as the principle of civil society and as moments in the constitution of the state, while others appear in the course of history, particularly the history of art, science and philosophy.[18]

The examples of 'subjective freedom' I wish to draw on here are precisely the 'other forms' mentioned here - the 'principles of civil society', and the freedom characteristic of the pursuit of science, art, philosophy etc.. Civil society as Hegel understands it has the duty to provide its members not simply with a 'place' in the form of work which earns subsistence, but work which is *meaningful* - not mindless toil in which the only reward is a wage, but activity in which the citizen perceives himself affirmed as an individual, and afforded recognition as a member of civil society. Such self-realising activity is to be contrasted with the fragmentation characteristic of what Marx was later to conceive as 'alienated labour', which it is illuminating to view in terms of the deficient form of the will outlined above, in terms of two 'moments': there is the first moment of 'freedom' as dis-identification with what I am doing, and there is the antithetical second moment of total bondage to an alien sphere of particular action that is, *ipso facto*, never actually taken up by me and owned as *mine*. The 'freedom' of the self thus remains wholly abstract, never *realised*.[19]

This deficient form of the will is further exemplified in the divided picture of the self and its moral nature that arises from certain readings of Kant's *Groundwork*. This text apparently leaves us with no way of accommodating Hegel's 'right of subjective freedom'. The requirements of duty and the categorical imperative on the one hand stand in opposition to 'impulse' and 'inclination' on the other, and neither category is adequate to what Hegel is trying to say. Hence if I find satisfaction and fulfilment in composing music or writing philosophy, this can only be viewed as 'inclination', and therefore as something which, in principle, I should be seeking to overcome. Ideally, on this picture of the human lot, I should aim to overcome such weaknesses as satisfaction and self-fulfilment, and pursue my musical or philosophical activities *purely* out of a sense of duty. Surely this is an absurd requirement, contrary to the wealth of human experience, and *belittles* all kinds of human achievement. In fact Hegel points out these consequences in the paragraph immediately following, where he depicts a conception of morality 'as nothing but a bitter, unending struggle against self-satisfaction, as the command: "Do with abhorrence what duty enjoins"', and then goes on to argue how this conception 'understands how to belittle and disparage all great deeds and great men'.[20]

Having said this, of course this is a very unsympathetic and one-sided reading of Kant. We need to redress the balance. A more sympathetic reading would be that the sense of self-fulfilment and satisfaction, while not something to be struggled with and overcome, is rather something to be tolerated, simply accepted as neutral, from an ethical point of view. But precisely what Hegel is saying here is that the sphere of realised subjective freedom is not 'neutral from an ethical point of view' as *he* understands the term 'ethical'. It constitutes the most concrete realisation of our freedom and of our capacities as moral beings. Hegel distinguishes 'morality' from what he calls 'Ethical Life' (the usual translation of 'Sittlichkeit'), reserving the latter term for 'actual' or 'realised' value - a sphere that encompasses the unspoken values embedded in love of another, in artistic creativity, in non-alienated labour, and so on. Hegel's point is precisely that this sphere exists *sui generis*, and that it has a 'right' against 'morality', the highest form of which is, for Hegel, the Kantian. Hegel is not so much *taking issue* with Kant's notion of morality - rather he is saying that it amounts to an incomplete standpoint of thought about the moral and the ethical.

At this stage we should recall the purpose of this detour into Hegel. The context was my argument, *contra* Sartre, that the fact that we are capable of sustaining 'constitutive attachments' etc., whatever these might be for each one of us, points towards what is perhaps *quintessentially* human, and is to be counted as a source of strength rather than weakness. This was part of my overall argument that Sartre's ontology of the subject cannot accommodate the wider conception of the will that he nonetheless explicitly avows. But the 'longer way' of locating constitutive attachments in a more comprehensive picture of the self is, as yet, only partially travelled. So far, I have commented only on the will. Much more needs to be said, for example about character, and more generally about the 'logical geography' of our concepts of selfhood and personhood. In the sections that follow, I hope to make good this promise.

I shall conclude this section with two questions. The first is this: How are we to characterise the *difference* between, say, a relationship with another person in which I feel myself affirmed and liberated, and one where I feel myself denied and constricted? Again, to take another example: How are we to characterise the difference between work that I find fulfilling, and in which I feel myself realised, affirmed and expanded, and work where I find myself frustrated, denied, constricted - or, in short, alienated? Is it *enough* to speak only of 'feelings' here - to say, in other words, that the difference is in the most basic sense 'purely subjective'? If we say this, then we seem to be endorsing a conception of human life in which every problem about our sense of identity, about what we as individuals should do, is to be resolved by the universal prescription of happiness drugs like prozac. So long as we *feel* alright, there is no problem. (It is tempting to add here that as noumenal selves we are untouched by such mere externalities as what we do in our everyday lives anyway.) It should be clear, both from the foregoing and from what follows, that I do not endorse any such conception of human life.

The background to my second question is this. Experiences of 'meaningful activity' or 'self-realising activity' are fairly commonplace - one does not need to be an artistic genius or prolific composer to experience such things. Nonetheless, such commonplace experiences do not fit into, hence cannot be accounted for on, the traditional theories of the self that constitute what I called, in the opening section of this chapter, 'the old agenda of the self'. In particular, they cannot be accounted for on the picture of the self that insists on a rigid and exclusive division between a sphere of 'reason' (as arbiter of 'duty') on the one hand, and the sphere of 'appetite' or 'inclination' on the other. But on what I have called 'the wider conception of the will', these commonplace experiences seem to be readily accommodated. Indeed Hegel claims, in his 'third moment' of the will and in his notion of 'the right of subjective freedom', to be providing just this accommodation. My second question is, acccordingly, this: How is this wider conception of the will possible, i.e. how (if at all) must our conception of the self be redrawn so as to accommodate it? (Allied to this there is a subsidiary question: Has Hegel provided the beginnings of an *answer* with his third moment of the will, and with his redrawing of 'subjective freedom', or has he merely renamed and further dramatised the problem?)

I address these questions in the course of the inquiries that follow, in my further pursuit of the 'longer way'.

6 Reinstating Character: a Preliminary Outline

Much of the foregoing, particularly in sections 4 and 5, points to the need to reinstate character at the centre of the self. But it must be clear from the course of this chapter that character cannot readily sit alongside the self as understood on 'the old agenda'. As we saw in section 4, viewing character as some kind of inner *object* I might discover, what I called 'an inner object of the second kind', does not give us anything worth having. It simply reinforces the notion of the subject that we need to get beyond. And viewing character as 'the subject' understood in the old way does not work either - for this notion of the subject is introduced precisely to fill the rôle of that which owns experiences, but can never *itself* be experienced or known in any way. If we are to take character seriously, and to seek to restore it to the centre of the self, then we have to abandon the old conception and redraw the self.[21] But how? The old conception is hard to dislodge.

To begin with, we need to clarify just what we mean by character, and what we are distinguishing it from. Of its doubtlessly many aspects, perhaps the most obvious lies in the common perception that character is relatively stable and unchanging. The sense in which this is so needs to be made clear.

There is some kind of distinction between one's character on the one hand, and such things as one's external appearance (e.g. facial features etc.) and one's habits (e.g. smoking, habitual bodily postures etc.) on the other. My facial features, for example, change only very slowly, unless for some reason I choose to have plastic surgery. If I did choose to have my facial features changed in this way, this would not normally make any difference to my character, unless my sense of my outer appearance played a dominant rôle in my life (as it often does with stars of the screen). Having said this, there are aspects of one's appearance that may be windows onto one's character (there comes to mind here the image of anxiety on the face of the young T.S. Eliot, in various photographs reproduced in various biographies). Similar things can be said about habits - I *can*, and frequently do, change my habits without this in any way affecting my character. But again, some habits - for example, smoking - may reach down much deeper into oneself, into one's character, and changing them may correspondingly be both more difficult and have an impact at that deeper level.

Thus while changes in one's appearance or in one's habits may be effected without causing great anguish, the fixity of one's character is, by contrast, of a different order. I may be advised or urged by my friends to change what they see as an undesirable feature of my character - 'don't be so jealous', 'try not to be so impulsive'. In such cases, some change in the desired direction may be possible. If I have, as we say, a jealous and possessive nature, I may become reflectively aware of this and begin to free myself to some degree from its hold over me. Such things are rarely achieved by mere 'introspection' - they usually require such intersubjective settings as, say, conversations with my close friends or my counsellor/psychotherapist. Over a period of time I may even cease to be jealous and possessive - I may come to 'understand' something with such a high degree of conviction that jealous and possessive behaviour of all kinds is no longer meaningful for me. On the other hand, I may well 'understand' that being jealous is irrational, of no avail to me, I may strongly disapprove of this aspect of my character, I may even make exertions in the form of 'acts of will' to resist my jealous and possessive tendency, and yet remain powerless to change it. What is the *difference* between the two cases? Why does 'understanding' in the first case facilitate self-transformation, whereas in the second it does not? Are there different *kinds* of understanding operating here? I shall return to these questions later. But first, further discussion of the unchanging nature of character is called for, in order to close this preliminary discussion of the topic.

Let us consider first the terms 'constitutive attachments' and 'categorical desires', which are owed respectively to Michael Sandel and Bernard Williams.[22] For Sandel, constitutive attachments are what define a 'thick' as opposed to a 'thin' conception of the self. What distinguishes a conception of the self as 'thin' is that it is 'given prior to its ends' - and this means that my character, my deepest convictions (moral, political, religious etc.) and my close ties to particular people (e.g. family, close friends) are all external objects of possession, definitive of what I *have* rather than what I *am*. Clearly the self understood in this way corresponds to those conceptions of the self, inherited from Descartes, Hume and Kant, that I am claiming we need to overcome. On the 'thick' conception of the self, by contrast, my character, deepest convictions and close personal ties define what I *am* rather than merely what I have - they define my 'constitutive attachments' in Sandel's sense. Clearly whatever my *particular* constitutive attachments may be, they will have to be relatively unchanging (my *capability* in this regard could not, of course, change at all). I could not 'choose' them in the way I might choose between different products in a supermarket, despite what may appear to be counter-examples. An entrepreneur, for example, may *appear* to be capable of changing his constitutive attachments 'at will', as we say, because of his rapidly changing life circumstances - but on more careful scrutiny, we see that these rapid changes in his external life exist at the service of his abiding entrepreneurial goal. The particular goals defined by these changing circumstances are conditional on one thing: whether they serve the ultimate and (in our example) *unconditional* goal of success in entrepreneurial matters - and here it goes without saying that one person's unconditional goal may be well down the scale of importance for another person, for whom a personal relationship might occupy that central place.

This notion of the unconditional leads on to Williams' 'categorical desires' or 'ground projects', notions which serve roughly the same purpose as Sandel's constitutive attachments. Williams brings out one central aspect of this unconditionality with particular force: our categorical desires are such that life would lack any meaning if fulfilling them became impossible; hence they provide the unspoken, unthought reason why we carry on living at all, and in Williams' discussion they need be neither selfish nor self-centred (our prime goal or ground project may be - to take two possible cases - leading a religious life of a certain kind, or being deeply concerned about the welfare of others).[23]

Finally, in order to complete this preliminary discussion, I return briefly to the will, which may seem antithetical to the very notion of character.   Later on in this chapter, I hope to show that, far from being antithetical, will and character are mutually implicated, the one illuminating the other.   But we need to see how character and will can come to *appear* antithetical.

I begin by invoking again the starting-point of Kant's *Groundwork*, where he opens with his famous statement that the only thing that is good without qualification is a good will, and then goes on to say that qualities of character and the like have only *conditional* worth.   In so far as character has only conditional worth, it belongs to the sensible world rather than to the intelligible.   The good will, indeed will itself, belongs to the intelligible world.   But since we are not morally perfect, our will may be 'pathologically affected' (but not 'pathologically necessitated') by sensuous motives.[24]   This may appear to legitimise (I put it no stronger than that) a picture of will and character as antithetical, with will having the power to coerce, and bring under control, the set of dispositions, motives etc. that constitute character, which set is then deemed to be *un*willed, *un*chosen. And it goes without saying that the conception of the will at work here is the narrow and not the wider conception.

Will and character thus appear antithetical only on a 'thin' conception of the self, the self 'given prior to its ends', and with the associated narrow conception of the will.   On a 'thick' conception of the self that allows for categorical desires, constitutive attachments etc., they are not antithetical, provided that we are prepared to endorse the wider conception of the will. On the 'thin' conception of the self and the narrow conception of the will, my character, my deepest convictions, my close personal ties etc. can be conceived only as objects of choice, a choice between a set of available alternatives.   On the 'thick' conception of the self and the wider conception of the will, they are objects of *discovery* rather than choice.   Clearly we need to look more closely at these terms, 'choice' and 'discovery', for on the face of it they seem mutually exclusive.   That they are not will, I hope, emerge in the course of the sections that follow, where one of my principal aims will be to exploit a certain conception of the self to be found in Plato's dialogues, most notably in the *Republic*.

7   The Tripartite Psyche in Plato's *Republic*

Although it occurs elsewhere in Plato's works (most notably in the *Timaeus* and the *Phaedrus*), it is in the *Republic* that we find the most complete exposition of a conception of the self that is both interesting in its own right, and interesting also in the fact that it has apparently had so little impact on the subsequent European philosophical tradition.   That conception is usually discussed only by philosophers, classicists and political theorists with an interest in Plato, and usually in the context of making sense of the theory of the ideal state (and related matters) to be found exclusively in the *Republic*. It is viewed by most writers in the field as a curiosity peculiar to Plato, and to have no wider philosophical relevance beyond exegesis of Plato. I refer, of course, to the theory of the tripartite division of the psyche. As I have argued elsewhere, we do have something to learn from it - it can inform our understanding of the self, and in particular, perennial and ongoing debates over the centrality of character, the nature of the will, and so on.[25]

Two of the three divisions are familiar to us, namely 'appetite', and 'reason' or 'intellect' (respectively translations of ἐπιθυμητικόν and λογιστικόν). The third element (θυμοειδές) is unfamiliar to philosophers not conversant with Plato, and hence remains unexploited by them. The epistemological tradition in European philosophy from Descartes onwards has not felt the need for a third element in the psyche, and it is therefore hard for us to take on board the idea that this third part, 'the spirited part', exists *sui generis*. Even philosophers with an interest in the *Republic* find it hard to take seriously, and are drawn back into talk of reason governing or ruling the appetites *simpliciter*, when strictly they should add, at the very least, 'through the mediation of spirit'. In what follows, I hope to make good the neglect - as I see it - of the tripartite psyche in general, and more particularly this third 'spirited' element, and to draw on it in order to inform our understanding of character, dispositions, and the will. But before embarking on this, I shall say something about the wide range of possible readings of the relevant parts of the *Republic*.

In the order of exposition in the text, the account of the tripartite psyche follows an account of the state, the 'polis', and its three divisions (in effect, three kinds of citizens, distinguished by their abilities and functions in the polis). This could be taken to mean that it is the polis that is fundamental, and hence that the divisions in the psyche are to be understood via metaphorical extension from the divisions in the polis. But then the question arises: Could it not be that, despite the order of exposition, it is *the psyche* that is fundamental, and the polis is to be understood via metaphorical extension? There is, accordingly, a rich set of possible readings concerning what Socrates says about the polis, the psyche, and the relation of analogy that ostensibly relates them. If we take the order of exposition literally and view the theory of the psyche as a metaphorical extension from the polis, we will be disposed to think of the psyche in terms of 'ruler' and 'ruled', or 'governance' and 'the governed'. Within this perspective, a variety of possibilities open up concerning the style or mode of governance, all the way from totalitarianism, via benign dictatorship to paternalism, and thence to something more consensual (such as Aristotle's 'polity' or disinterested rule by the many.)[26]

It is easy to take the view that the polis is primary if we take this section of the *Republic* (441a-444e) in isolation. The section begins with the description of the three 'classes' in the polis and their mutual relations - the institutional setup, the duties of the rulers etc. - followed by the account of the four cardinal virtues (wisdom, courage, discipline and justice) and where (i.e. in which class or classes) each virtue is to be found. The three classes are not socio-economic, but defined in terms of their *function* in the polis. The guardians (φύλακες) are (on a more liberal reading) the source of understanding and wisdom, or (on a totalitarian reading) the ruling ideologues, those who in the end give the orders. The auxiliaries (ἐπίκουροι) translate and realise that understanding and wisdom (again on a liberal reading) into stable, workable institutions that meet the needs of all citizens, or (on a totalitarian reading) carry out the orders of the guardians. The third class are the ordinary citizens who engage in all kinds of productive, commercial, professional and artistic activity. Taking the liberal reading, we have in the guardians the wisdom (σοφία) of the polis, and in the auxiliaries the solidity, the steadfastness, in short the *courage* (ἀνδρεία) of the polis - and we can understand this in terms of institutions that are stable and not easily swept aside, because they are based on wise principles.

Only after the discussion of the polis is completed are we *explicitly* introduced to the threefold division of the psyche. First, reason or intellect is distinguished from appetite, and then the other element, 'spirit' (θυμοειδές), is distinguished from appetite by means of the famous example of Leontius' simultaneous fascination with the dead bodies outside the city wall, and his repugnance at this primitive urge - the former due to appetite, the latter due to spirit (439e-440a). Such spiritedness, when there is inner conflict, always comes in on the side of reason and against appetite - and it is distinguished from reason by the common perception of that spiritedness in children that often manifests in opposition to reason. Spirit is here characterised as 'by nature the auxiliary of reason' (ἐπίκουρον ὂν τῷ λογιστικῷ φύσει, 441a).

This, then, is how the tripartite psyche is *explicitly* introduced. What is curious about this discussion is how contrived it appears, once we realise the extent to which earlier parts of the *Republic* have already set the stage for it. Indeed it can be argued that the harmonisation of spirit and intellect is as much a central theme of the *Republic* as that of the harmonisation of political power and philosophic wisdom, *at least* from the discussion of the earlier stages of the process of the education of the guardians in books two and three onwards, if not earlier.[27] All of this suggests that it is the psyche with its divisions that is primary, and the polis with its three classes the metaphorical extension. Nonetheless it is sometimes illuminating to view the metaphor as going from the polis to the psyche, as when the spirited part is described as 'by nature the auxiliary of reason'.

Just as the explicit theory of the tripartite psyche is prefigured earlier in the *Republic*, so is the connection between the spirited element and character. The passages I shall focus on here are 375-6 and 410c-412b. The former (375-6) is where the guardian is compared to a good watchdog, and where we are told that the psyche of the guardian can manifest courage only if it is spirited (θυμοειδῆ, 375b). The point is also made in this passage (at 375e-376c) that the element of spirit must be tempered by reason. In the second passage (410c-412b), there is a discussion of how the combination of artistic or 'musical' education (μουσική) with physical training (γυμναστική) 'for the spirited principle and the philosophical' (ἐπὶ τὸ θυμοειδὲς καὶ τὸ φιλόσοφον) brings together and *harmonises* these two principles. All this prepares the ground for the book four discussion of the four cardinal virtues in the psyche. Courage is the characteristic virtue of the spirited element, and in the book four discussion of that virtue the courageous, steadfast soul is compared to a well-prepared fabric that is able to absorb and retain the dye - just as the colour remains fast in the rightly prepared fabric, so the psyche possessed of courage will be able to retain its right judgement, through the joys, sufferings, distractions etc. to which life subjects it, about what is and is not to be feared, and to act appropriately in accordance with that judgement.

Apart from the introduction of the 'spirited' element, the account of the tripartite psyche is pervaded by musical metaphors - harmony, attunement and so on. Where certain readings of the *Republic* find only the language of mastery, domination etc., Plato presents us with the recurrent theme of how intellect and spirit are to be brought into *harmony*, how the spirit is to be *tuned* to respond to intellect, how political power and philosophical wisdom need to be *harmonised*, and how the three parts of the psyche are to be brought into *harmony*.

We embarked on this discussion of the tripartite psyche principally in order to draw on it for a way out of the *impasse* posed by a certain conception of the self we discussed in section 4, namely the self as understood on the S→O picture. The question for character is this: How can we locate character in the self in a way that makes of it neither an inescapable object of direct acquaintance, nor something in principle unknowable because identical with S, the knowing subject? It will become clear in what follows that the spirited element of the psyche forms a central part of the answer to this question. But first of all we need to reflect on some radical differences between Greek conceptions of character (as found in Plato's *Republic* and Aristotle's *Nicomachean Ethics*, for example) and our own.

The account to be found in the *Republic* is normative, and it involves a typology - there are *types* of character. These two aspects are closely related. The central feature of the typology is that every character-type pursues 'the good', but in each case in accordance with the central disposition or 'ruling passion' characteristic of the type. The major typology is to be found in books eight and nine, which contain a scheme of five types: one ideal and perfectly balanced, four unregenerate and imperfect. Each of these perceives 'the good' in accordance with its nature. There is *one* good, 'the good itself', αὐτὸ τὸ ἀγαθόν - perceived as it is, or at least not misperceived, by the perfectly balanced psyche, and misperceived in progressively more distorted ways by each of the four unregenerate types. The five types are expressed in political terms: we have the balanced psyche corresponding to the ideal state or 'aristocracy', followed by the timarchic or timocratic psyche, then the oligarchic, then the democratic, then the tyrannical. The timarchic type misperceives the good as honour - he will be loyal to his superiors, tough on those over whom he has authority, ambitious to hold office or to prove himself in battle, of limited vision, distrustful of reason, and so on. This means that if there is a conflict between a more comprehensive vision of the good and his conception of honour, he will follow the latter. Similarly, he will disdain wealth if its pursuit offends his sense of honour. The next type, the oligarchic, has a more distorted misperception of the good. He misperceives it as wealth and social position, and so if for him there is a conflict between doing the honourable thing and pursuing wealth, he will pursue wealth. The democratic psyche and the tyrannical psyche are characterised respectively as having progressively more distorted misperceptions of the good. It would lie outside the scope of this essay to go into this in any detail, and so I shall say nothing more about the democratic and the tyrannical types.[28]

The central point for present purposes is the way the different types map onto the tripartite psyche. The timarchic character's strong yet narrow sense of honour, his distrust of the intellectual, his lack of vision - all these aspects of his being spring from an underlying balance of elements in the psyche in which intellect does not so much *inform* the spirited nature, but rather works as a mere instrument *used by* the spirited nature. This contrasts with the ideally balanced psyche, where the spirited element is *responsive* to the insights of the intellectual or reasoning element rather than either being dismissive of them or using them for its own purposes. The oligarchic character, on the other hand, is dominated by his desire to fulfil his appetitive needs, to acquire wealth, to acquire social position based on wealth, and so on. For him, therefore, both reason and the spirited element exist in the service of appetite - reason calculates how best to pursue his overriding aim, and the spirited element provides the energy to bring it to realisation.

The psychology of the *Republic* may seem too distant from our own notions of character to be of any use. It involves what appears to be a rigid typology, and it is normative in a way that is foreign to our own more complex situation. Nonetheless, it can help to illuminate many of the aspects of the self I have discussed in this chapter, such as constitutive attachment, categorical desire, and the wider conception of the will. To be the kind of person who, say, places the pursuit of honour above the pursuit of wealth is already to have grown into a value or set of values, and *ipso facto* to have 'willed' those values, in the wider sense of 'will'. It is to have, already, a 'conception' of one's own life as having an unquestionable basis, a basis on which one's particular projects and choices are grounded. It is to have a 'conception' of one's life such that, were its realisation somehow made impossible, one would have no reason to want to carry on in life at all. I place the word 'conception' in quotation marks to indicate that this ground project, this set of values assented to, does not have to be known to me with any degree of clarity. It is something pre-reflectively given, and does not have to be an object of knowledge at all. It is clearly something I *enact* in my various life-plans and projects, but not directly. I pursue my particular ends in a way that may be conditioned by this sphere of pre-reflective values, but I may be only *dimly* aware of this sphere, if at all. Hence the language of discovery has as much a place as that of enactment. The embedded values such as honour that serve to distinguish one character-type from another in books eight and nine of the *Republic*, despite the fact that they are *types*, can be viewed as categorical desires or ground projects in Williams' sense

At this point it would be appropriate to take a wider perspective on the three elements in the psyche. We can view the appetitive part as encompassing the wish to possess, as the source of needs and wants, and even as openness to or appetite for experience. We can view the spirited part as the source of the wish to be or to prevail, in a way that invites comparison with Spinoza's use of the term *conatus*, translated by the term 'endeavour': 'The endeavour wherewith a thing endeavours to persist in its being is nothing else than the actual essence of that thing'.[29] This perhaps calls for some explanation. As Stuart Hampshire points out, in Spinoza's system the 'actual essence' of each thing is nothing other than its tendency to maintain itself as the thing it is. 'The greater the power of self-maintainance of the particular thing in the face of external causes, the greater reality it has, and the more clearly it can be distinguished as having a definite nature and individuality.'[30] This last point can help us understand how the spirited part can attach itself to the pursuit of the good in and for itself, or the pursuit of the good misperceived as honour, or the pursuit of the good misperceived as wealth and material possessions, and so on, whilst at the same time explaining *how* the account of character in these parts of the *Republic* can be normative. Clearly if my understanding of myself goes no further than Plato's oligarchic type, I will view my wealth etc. as somehow definitive of me, and will therefore view my life without that wealth as not worth preserving. As the timarchic character, however, I would have no such attachment to wealth - my self perception in terms of honour would be less subject to external contingencies, and so I would have 'greater reality' in Spinoza's sense. If my *amour propre* is not tied to such externalities, there is no risk that I will perish when they do. Finally, the intellectual part of the psyche is the source of the wish to understand, to know what is real and what is not. It does not, considered in itself, have any power. To be effective, to bring about any changes, it needs to enlist or awaken the spirited part. I offer in what follows some examples of how 'awakening the spirited part' might be given an interpretation.

Let us recall the comments I made in the last section on two cases of self-understanding. In the first, my 'coming to understand' that I have a jealous and possessive nature is effective in bringing about a change in me - I become, over a period of time, less possessive. In the second, I also 'come to understand' that I have this undesirable trait, I may disapprove of it, I may even struggle and engage in all manner of 'acts of will' to resist it, but all to no avail. What is the difference between these two familiar scenarios? In order to elucidate this, I consider a recent conversation with a man who gave up smoking. I use this example because it illustrates both cases - initial failure, followed later by success. He was well acquainted with the medical evidence, had read of people giving up smoking, and knew of the various ways others had tried to give up. He tried to give up, used all the suggested ways of giving up, struggled to break the habit making appropriate 'efforts of will', but found himself continually slipping back into it. Then, on one particular occasion, he was reading something about smoking that was quite unremarkable, in that it contained nothing he did not know already - but in that moment, he underwent a change which he described to me in the words 'I no longer saw myself as a smoker'. Afterwards, he found it quite easy to dismiss the impulse to smoke, without having to resort to efforts, strainings or episodic 'acts of will' of any kind.

I want to suggest about this example that, in the later phase, what we have in terms of the tripartite psyche is intellect awakening spirit. We could also say that in this later, successful phase we have a kind of understanding that is not mere ratiocination, not mere information, but engages the depths of the self. Something that was previously mere information is now thought through with the whole of one's being. Notice, what the man says is not: 'I became convinced beyond all doubt that smoking was wrong'; or, even, 'that it was wrong for *me*'. What he says is: 'I no longer *saw myself as* a smoker'. The idea of awakening the spirited nature can be expressed in terms also of the will. What we have is will in the wider sense, rather than the narrow sense which in this case produced nothing but vain and ineffectual 'acts of will'. Again, we can say here that the self is engaged at the level of one's ground-projects or categorical desires - smoking becomes simply peripheral, and therefore easy to dismiss.

This example, trivial though it may be, may nonetheless show us something about what more important decisions can be like, namely that what we call 'the moment of decision' is often merely the making explicit of what has been already decided long ago. This is better illustrated by another example: that of the emerging and fateful relation between Cromwell and Richard Rich, in Robert Bolt's play *A Man for All Seasons*. Richard Rich is portrayed as a young academic who, finding himself in the world of public affairs, is full of self-doubt and wants to be rescued from the temptations of public life. But the way he wants to be rescued is for someone - initially, Sir Thomas More - to provide him with some sufficiently respected public position that will meet his hitherto suppressed desire for worldly success of this kind, but without his having to compromise himself. More offers him a schoolmaster's post, and in the face of Rich's evident disappointment, adds the perceptive comment that 'a man should go where he won't be tempted'.[31] Rich turns the offer down. In a later scene, Rich behaves in an uneasy way in front of More and Roper, telling them that Cromwell is 'collecting information' about More. (Rich has even at this stage been consorting with Cromwell, who has been advising him to read Machiavelli.) More is unmoved, and Rich begins to tremble and beseech More to help him by employing him, because he is 'adrift'. More refuses.[32] There is then a meeting between Cromwell and Rich, where Cromwell gets Rich to admit, with some bitterness, that he would pass on information if he was offered enough for it. When Cromwell remarks how depressed he looks, Rich replies, with mock light-heartedness, that he is lamenting the loss of his innocence. To this Cromwell replies 'you lost that some time ago. If you've only just noticed, it can't have been very important to you'.[33] Now that he has become Cromwell's protégé, all that he needs to do in order to betray More (who will not acknowledge Henry VIII's right to divorce) is to square it with his conscience, and here Cromwell eases the way for him with such phrases as 'administrative convenience'.

My point in citing this example is simply to illustrate how a decision about one's priorities can be taken almost unnoticably, long before one has made it explicit, even to oneself. The example also illustrates further the appropriateness of the wider rather than the narrow conception of the will. Clearly Rich does not suddenly decide, in the way we might decide between watching channel three rather than channel four, that some ground project *A* matters more to him than something else, *B*. It was as much a process of self-*discovery* as one of self-*enactment* that Rich found - not without some initial bitterness - that he was corruptible, to the extent that in the end he was able, on oath, to bear false witness against Thomas More. This process of self-discovery is made bearable to the extent that his political career takes the ascendency - and to the extent that this happens, conscience is silenced and placed beyond reach.

I want to end this discussion of different kinds of self-understanding with some further reflections on the rôle of thinking. What the smoking example shows is a kind of thinking, best characterised as a dawning realisation in our terms, but which in terms of the tripartite psyche is a clear example of reason ruling appetite through the mediation of spirit, its 'natural auxiliary'. We might see this in another way, namely that the appetite for smoking is finally put in its place by what we might call conviction, a spirited sense of who one is that excludes smoking as either a necessary or even a remotely useful aspect of who one is, informed by an understanding or reason that sees and understands *in general* how pointless smoking is. One can therefore understand those readings of the *Republic* that view the relation between intellect and spirit, and in turn between spirit and appetite, as *consensual* rather than dominating.

What the example of the fateful relationship of Rich and Cromwell shows is something interestingly different. In terms of the tripartite psyche, we have in Rich a bright intellect alongside a weakness and irresoluteness of spirit, in turn attached to suppressed appetites for wealth and social position. He gains in such 'strength' as it is open to him to have, but only in proportion to how far he ascends the ladder of his political career - spirit and intellect are mere servants of his unbounded ambition for social status. It is not without some initial bitterness that he comes to see these things about himself. His only hope for peace is for him to see them less and less - a state he achieves at the end of the play. A *prudent* Rich, made wise to these things earlier on in his life, would have been better off as a schoolmaster.

8   The Self and its 'Logical Geography'

The tripartite psyche thus emerges as a combination of (a) desire, need and openness to experience, with (b) the wish to be and to prevail, and further, (c) the wish to understand what truly is. Towards the end of book nine of the *Republic*, the appetitive part is likened to a many-headed beast, the spirited part to a lion, and the intellectual part to a man.[34] The plurality of conflicting wants symbolised by the many-headed beast can be seen as confronting the will with the threat of conflict and fragmentation, with the lion of the spirit (informed by the man, the intellectual part) harmonising them into a coherent whole. This complex image should be viewed as a summary rather than a substitute for book four and the discussions leading up to it - for after all it is at the end of *book nine* that we are given it. It is, nonetheless, a useful addition. Be that as it may, the picture of character that emerges from the tripartite psyche may seem a far cry from our own understanding of character. What we should notice here is how, as a picture of the self, this contrasts strongly with the picture that comes to us through Rousseau and Kant, in which a part of the self rules and another part is 'brought to heel'. We have to leave this picture behind because it cannot accommodate categorical desires, constitutive attachments, character and the like. It is absurd to hope that we can account for these things through conceptions of the self that require one or other of two conceptions of the moral or the ethical: one based on an impersonal calculation of utilities, the other on an impartial application of universal maxims. If these ways of thinking about the ethical are to have any place, it must be because of the prior claim that lies in the life of a person, replete with his or her loves, aspirations, convictions etc.. These ways of thinking can certainly *qualify* this prior claim, but should never be thought of as capable of replacing it. (The basis for this conclusion lies in section 5, in the extended discussion of the will.)

Those philosophers who seek to endorse a 'thick' rather than a 'thin' conception of the self tend to emphasise the self's *social* character, with very little by way of explanation as to how this is possible - the individual self becomes, somehow, a rag-bag of elements simply absorbed from the community or society in which it lives. Even Hegel, with the notion of 'subjective freedom' referred to earlier, and with his discussion of the will in its 'third moment', goes only part of the way towards redrawing the self in a manner that would make the realisation of that freedom comprehensible. It will be recalled that we left open the question whether Hegel succeeds in redrawing the self, or merely re-states and dramatises the problem

I am suggesting that we should view Hegel's discussion of 'the right of subjective freedom', of the third moment of the will, and everything that derives therefrom, as addressing *the same issue* as is addressed in the *Republic* through the intermediate, spirited element of the psyche. These notions of Hegel, culminating in the discussion of 'Ethical Life in its immediate phase', constitute (among other things) his attempt to break free from the Cartesian inheritance and to carry out his own programme of redrawing the self. It is always possible to ask how successful Hegel was - or indeed how successful more recent philosophers such as Strawson or Merleau-Ponty have been - in freeing us from the Cartesian picture, or from what I have called 'the old agenda of the self'. These old pictures of the self are always there, sedimented in the way way we think and speak. The task of redrawing the self can never be accomplished if what we are aiming for is a new conception with *that* degree of sedimentation. The tripartite psyche in the *Republic* may well have been, or have developed out of, the sedimented folk psychology of its times. Hence it would be a mistake to try simply to *graft it on* to contemporary critiques of the 'thin' conception of the self, such as Sandel's, that I have drawn on in earlier sections of this chapter. At best, all we can hope for is that the rôle that the spirited element plays in the *Republic* can begin to *inform* these contemporary critiques.

Notes and References

[1] However, the transcendental argument for self identity does not have to proceed in this way. Strawson has argued that there is a logically non-transferable sense of the words 'my' and 'mine' which is ineliminable, and which does not appear to have any transcendental idealist implications. Strawson appeals to this sense as part of his own argument for the necessity of acknowledging the logical priority of the concept of a person over such concepts as 'ego', 'self', 'mind' and 'body'. See P.F. Strawson, *Individuals*, London, 1959, chapter 3 *passim*.

[2] M. Merleau-Ponty, *Phenomenology of Perception*, (trans. Colin Smith), London, 1962, p. 26 ff.

[3] Immanuel Kant, *Critique of Pure Reason*, (trans. Norman Kemp-Smith), London, 1961, p. 152.

[4] Immanuel Kant, *Groundwork of the Metaphysic of Morals* in H.J. Paton (trans.), *The Moral Law*, London, 1972, pp. 116-119. In the rest of this chapter I refer to this text as the *Groundwork*.

[5] Kant, *Groundwork*, p. 62.

[6] Ibid. p. 62.

[7] The term 'logical geography' is Gilbert Ryle's. See Gilbert Ryle, *The Concept of Mind*, London, 1949, Introduction and passim..

[8] Sydney Shoemaker, *Self-Knowledge and Self-Identity*, Ithaca, New York, 1963, p. 65 and chapter 2 passim..

[9] Leo Tolstoy, *Anna Karenin*, trans. Rosemary Edmonds, Harmondsworth, 1972, p. 118.

[10] Plato, *Republic*, trans. Paul Shorey, Cambridge Massachusetts, and London, 1978, volume 1, 435D. The Shorey translation speaks of 'a longer and harder way'.

[11] Andros Loizou, 'Scepticism, Embodied Existence and Time', *Skepsis* VIII, Athens and Olympia, 1997, pp. 194-204.

[12] See Spinoza, *Ethics*, trans. Andrew Boyle, London and New York 1959, Part 2, Proposition 49, corollary; and Hegel, *Philosophy of Right*, trans. T.M. Knox, Oxford, 1967, p. 21.

[13] J-P. Sartre, *Being and Nothingness*, trans. Hazel Barnes, London, 1966, p. 142.

[14] Hegel, *Philosophy of Right*, p. 23.

[15] Ibid., p. 24.

[16] Ibid., p. 228 (my italics).

[17] Ibid., pp. 261-2.

[18] Ibid., p. 84.

[19] For Marx's characterisation of alienated labour see David McLellan (ed.), *Karl Marx: Early Texts*, Oxford, 1979, p. 133ff.

[20] Hegel, *Philosophy of Right*, p. 84.

[21] The reader may have noticed that I speak of *restoring* character to the centre of the self, thus suggesting that it once enjoyed that status, or that in the common understanding, i.e. among ordinary folk innocent of philosophy, it enjoys that status. As will emerge presently, there is truth in both assertions.

[22] See Michael Sandel, *Liberalism and the Limits of Justice*, Cambridge, 1982, p. 179 and passim; and Bernard Williams, 'Persons, Character and Morality', in A.O. Rorty (ed.), *The Identities of Persons*, Berkeley, Los Angeles, London, 1976, p. 207.

[23] See Williams, 'Persons, Character and Morality', pp. 207-210.

[24] See Immanuel Kant, *Critique of Pure Reason* (trans. Norman Kemp Smith), London, 1961, p. 465.

[25] See 'Eikasia and Tyranny in Plato's *Republic*', in Andros Loizou and Harry Lesser (eds.), *Polis and Politics: Essays in Greek Moral and Political Philosophy*, Aldershot (UK) and Vermont (USA), 1990, and also 'The Threefold Psyche and the Dramatisation of Justice in Plato's *Republic*', in *Polis*, Volume 16, 1999.

[26] For a balanced discussion of totalitarian readings, see C.C.W. Taylor, 'Plato's Totalitarianism', *Polis*, Volume 5, No. 2, 1986. In the end, Taylor opts for a paternalistic reading. A more consensual reading is *strongly suggested* by N.J.H. Dent, 'Plato and Social Justice', in Andros Loizou and Harry Lesser (eds.), *Polis and Politics*. See also R.W. Hall's papers, 'Platonic Justice and the *Republic*', *Polis*, Volume 6 No. 2, 1987, and 'Plato and Totalitarianism', *Polis*, Volume 7 No. 2, 1988. The term 'polity' first occurs in Aristotle, *Politics*, 1279a35-40, where it is defined as 'rule by the many in the common interest'. (This is later distinguished from 'democracy', which is defined as 'rule by the many in their own interest'.)

[27] I argue the case for this, among other things, in 'The Threefold Psyche and the Dramatisation of Justice in Plato's *Republic*', *Polis*, Volume 16, 1999.

[28] The reader is referred to Harry Lesser, 'Society and Individual Psychology in Plato's *Republic*', in Loizou and Lesser (eds.), *Polis and Politics*, for a full and interesting discussion of these issues.

[29] Spinoza, *Ethics*, Part 3, Proposition 7, p. 91.

[30] Stuart Hampshire, *Spinoza*, Harmondsworth 1951, p. 122.

[31] Robert Bolt, *A Man For All Seasons*, London 1960, p. 3.

[32] Ibid., pp. 35-36.

[33] Ibid., pp. 40-41.

[34] Plato, *Republic*, 588c-589a.

4 The Time of the Embodied Self

1 Introduction

Time as we have considered it so far, i.e. in the first two chapters, has been the time that presents itself to consciousness as an object, i.e. time *for* the self. At the end of chapter two, we sketched an account of what I called 'the time of our plans and aspirations, our achievements and failures, the stories we tell ourselves of our origins', in which some new terms were introduced - unsaturatedness, significance, emerging identity, to which we added the more familiar term 'narrative'.[1] But this remains an account in which the narrator is in an important sense outside his narrative, despite the fact that he may be one of its *dramatis personae*. In so far as he tells - or writes - a story in which he is a player, in so far as he views himself as enclosed within his narrative, he views himself as just a player alongside the others; he views himself as another. The strain begins to show once our narrators begin to refer to themselves in the first person, once they say 'I'. For then they stand both within and outside the time of their narrative. The grandmother telling stories of her childhood to her grandchildren may add 'that young girl who joined the volunteers was me'. By this statement, she includes her listeners, her grandchildren, in a wider story that is still open, in which they too are participants. Of course this transition from the third person, singular or plural, to the first person, singular or plural, is latent even in the grandmother's recounting of a story told her by another in which she was not a participant, and therefore in which her grandchildren can have no direct personal stake. But surely even here her young listeners must conceive of the unknown players in the recounted story as beings who are capable of saying 'I' and who are, like themselves, both observers and participants in the story of their lives. What we need to do next is to seek an understanding of how it is possible that we can be both observers *of* time and participants or agents engaged *in* it.

Following the extended argument of chapter three, we no longer have to be dictated to by the 'thin' theory of the self summarised in the S→O schema, i.e. we need no longer see the subject as either wholly outside the time of its experiences in accordance with the schema, or totally immersed in time as an anonymous series of events if we are moved to deny the schema for Humean reasons. The alternative conception that emerges in chapter three is one in which the self is grounded in its prereflective life. It matters little whether we express this in terms of 'ground projects', 'categorical desires', 'constitutive attachments', 'the centrality of character', 'dispositions' or 'will in the wider sense', or in terms of the tripartite psyche of Plato's *Republic* suitably interpreted - for as became clear at the end of chapter three, the notions that these terms answer to exhibit a kind of confluence in which each one of them is capable of illuminating the others. In what follows I shall freely deploy whichever of these notions seems most appropriate.

The question that constitutes the chief concern of this chapter is this: How are we to understand the relation between time and the self - or again, in what way is the self temporal? I shall now begin to address this question.

2 'Ground Projects' and their Temporality

Much of the last chapter was devoted to giving the self 'substance' in an acceptable sense of the term, i.e. without invoking the category of that name; to have done otherwise would have been to go against Hume's valid point that the self is not an introspectible *object*. The self cannot be a mere sum of its experiences, nor can it be any introspectively discernible principle uniting them. But neither can it be wholly outside time. The alternative conception of the self advocated in chapter three cannot be thought through in either of these ways. It does not simply *persist through* time, but neither does it *stand outside* time. We are in need of an alternative.

I begin with the discussion of the extent of the present in section 5 of chapter two. The point I focus on is the analysis of propositions in the continuous present tense - 'it is raining', 'John is walking his dog' etc.. The logic of such statements requires that the event or process in question has begun and not stopped, and therefore that it is *legitimate* - given the other supporting arguments - to speak of past and future parts of the present event or process. The point can then be made that there is no *logical* reason why we should not extend our use of the continuous present tense - as in fact we do - to say such things as 'X is writing a new English dictionary', 'Y is running a political campaign' and so on. This seems to entail that events or processes of *any* duration can be present. The only objection that might be brought against this is a psychological one, or possibly an epistemological one. There is no way that, say, 'the ozone layer is getting thinner' can be given a direct psychological meaning. But having said this, there are other reasons for us to be sceptical about the notion of a psychological present anyway. Does it refer exclusively to perceptual experience? There is no reason why it should. My present action can just as legitimately be considered to be 'psychologically' present as my present perceptual experience. But then, what is to count as 'psychologically present' in the sphere of action? What is to be the criterion for deciding what falls within it? Mere duration will not do. 'Significance to the agent' sounds more promising, but threatens to undermine the whole notion. The fact that I am engaged on a particular long-term project may be more significant to me than the action I am engaged in at this moment. To take an example from the last chapter, Richard Rich's ground project is such that his personal allegiances are almost exclusively determined by his need for a certain kind of recognition and personal advancement, and this is so without his having 'consciously' to sustain it with 'acts of will' understood in the narrow sense. The ground project itself is what remains, and in relation to it this rather than that relationship to another person, this rather than that particular value, this rather than that commitment, are one and all expendable if they do not serve the ground project.

We need to look closely at what constitutes ground projects and categorical desires. The terms were introduced in order to accommodate the fact that each person will have certain aims, attachments, aspirations - not necessarily *named* as such, and of course different for each person - which provide for that person some reason for wanting to carry on living at all. How are these different from long-term projects, like the project of re-translating all of Plato, or the long-term leadership of a political movement? There are a number of crucial differences. The long-term project may be decided upon after deliberation, in which one may put to oneself questions like 'Do I really *want* to spend a large part of my life re-translating Plato?' What tips the balance will certainly have a great deal to do with my 'ground projects' or 'categorical desires'. These, by contrast, are not 'chosen' in the way choice is ordinarily understood, viz. through what I called in the last chapter 'will in the narrow sense'. Nonetheless, my long-term projects are likely to be chosen in a way that realises these ground projects. The second crucial difference between long-term and ground projects is that the latter need not be known to me with any degree of clarity. My long-term projects, by contrast, belong in the sphere of conscious purpose in the everyday sense, and so the question of 'knowing' about them does not arise. A third difference has to do with time-scale. I know when something I have elected to do, such as translating Plato or being involved in a political movement, begins and ends. A ground project, by contrast, emerges from the unknown depths of my past into some kind of penumbral shadow, it has no definite temporal starting-point, and it gives an imperceptible shape to my future, laying out in advance the possibilities open to me. It is important to see this as not merely negative. To enclose possibilities within boundaries is to give them a certain solidity which they would otherwise lack; it is to transform abstract freedom into concrete freedom, freedom in relation to the real situation of the agent, freedom *within* his or her world.

Richard Rich's ground project reveals itself in the course of the play, not only to the reader or the audience, but also to Rich himself. As the fateful events take their course, as Rich both enacts and discovers who he is, so there is no turning back for him - the future closes and solidifies round certain possibilities, namely those to do with personal advancement, to the progressively greater exclusion of anything that might speak to higher ideals. He is a failure by the highest standards, but he has the success that he really wants. On the other hand Pasha Antipov ('Strelnikov') in Pasternak's *Doctor Zhivago* is by contrast a man of high ideals, ideals that qualify his love for Lara. He is married to her, they have a young daughter, yet in his lofty way he feels unworthy of her. He feels the need to do great deeds to make himself worthy of her love, enlists in the 1914 War, then moves on to espouse the Bolshevik Revolution, which he then serves for a number of years with zeal. Throughout these years, first in the War and then in the service of the 1917 Revolution, he is parted from Lara and his daughter. One day, he tells himself, he will return to them - but for him they belong to another life, a life that will be his only when he has seen *this* one through, when he has achieved whatever he needs to achieve in the name of the cause. When, in the end, it becomes clear that the cause has turned against him, and that he will never see Lara again, when he realises that as a result of this double loss life no longer holds anything for him, he shoots himself. He preserves his ideals, and judged by certain very high standards his life is not a failure - he is not prepared to settle for less, to the extent of paying with his own life. Within the range of what is possible for him to have, he may not get what ultimately he would like, but arguably he gets the next best thing in that he avoids the worse scenario of being hunted down by those who, in his eyes, have compromised the revolutionary cause.

3 Sartre and Ecstatic Temporality

It has seemed, so far, that either the self is *in* time and persists or endures *through* time, or it is outside time altogether. In contrast to these positions, what I shall argue is that the self is temporal in a more radical sense. It *is* its time, and it is that which gives *structure* to its time. But how are we to understand the principle that unites the self and gives it its structure? It can be neither outside time, nor simply the events or experiences it unites. It must be an immanent, indwelling principle that somehow produces *itself* as well as its time. Sartre's account of consciousness as being for-itself can be seen as an attempt, if not to provide such an immanent principle, at least to address the issue that seems to require such a principle. The reason why I hesitate in calling Sartre's being for-itself and its acts an immanent or an indwelling principle will emerge in what follows.

Let us now return to ground projects. My ground projects are the *conatus* that gives me a stake in the world, propelling me forward into it and providing me with my own real possibilities, but they also constitute the limits of my world. This dual character - real possibilities, real limitations - also qualifies these ground projects in their temporal aspect.

As I face the everyday contingencies of my life, the choices I make are not in a vacuum but within some framework that my past being, and not just my past choices, has already in part determined. My ground projects thus form a kind of temporal horizon that encompasses my past being, my present, and whatever real possibilities are open to me in the future. Pasha Antipov's love for Lara, together with his lofty idealism, are not elements or 'inner objects' that he introspects, but rather his very *being* - not something he *has*, but something he *is*. But on Sartre's account of consciousness this is inexplicable, as we saw in chapter three - for on that account, they would have to be like 'inner objects', in so far as they would be qualities discernible in past acts in relation to which his consciousness, as being for-itself, is a detached onlooker.

Nonetheless, there is something of positive value in Sartre's account. This is because his conception of being for-itself is necessarily temporal in a way that the theories of 'the old agenda' that I sought to overcome in chapter three are not. A purely timeless self, such as Kant's noumenal self, could not *ex hypothesi* be temporal in the sense required - its *experiences* would be in time, but would be merely objects of passive witnessing. The only relation between a timeless self and its 'time' would be one of total disidentification. As we shall see, Sartre's account of the self's temporality goes a step beyond this in having, in-built, at least the *striving towards* some kind of substantiality in the world, some kind of temporal being - even if, in the end, it can never embrace its experiences. A self that was nothing more than the series of its experiences would consist simply of the arrival of one experience after another with no grasp of its situation, since on such a view any kind of reflective consciousness would be impossible. Sartre's theory clearly avoids this pitfall too.

Let us pass on, then, to the detail. The terms 'being in-itself' and 'being for-itself' are correlatives in Sartre's *Being and Nothingness*, and so the sense of one presupposes, and is presupposed by, the other. But in the Introduction, Sartre examines being in-itself on its own, for what must be purely expository reasons. Were such a state of affairs possible, 'pure being in-itself' would have to be characterised by a plenitude of such a kind and in such a way that there would be no room for the possible - no gaps, no element of 'non-being' which would allow possibility, or movement, or change. It would have to be beyond all characterisations other than plenit- ude; beyond all distinctions of active and passive, beyond all differentiations between one feature and another, for such distinctions or differentiations pre-suppose or invoke meaning, purpose etc.

Consciousness as 'being for-itself' is the complete opposite of being in-itself thus characterised. It is a vacancy, an absence, a translucent mode of being, one characterised by a distancing - even a distancing from *itself*, should it ever find itself infected with anything of a determinate character. It is a sphere of being where alternative possibility is inescapable. Being in-itself, in the manner indicated in the last paragraph, is by contrast too dense, too solid, for there to be alternative possibilities. As Sartre notes at the end of the Introduction, the two spheres of being thus described are mutually exclusive.[2] Yet it is essential to Sartre's philosophy that they be inescapably bound together. In what is in many ways a typically Hegelian move, Sartre characterises the mutual exclusivity of the two spheres so far adumbrated as 'an abstraction', and sets about providing his own phenomenological descriptions of instances where the two modes of being are shown in necessary relation. The parallel with Hegel consists in the way these instances of necessary relation describe a reality in relation to which the two modes of being are exhibited as subsidiary moments - a reality which therefore *grounds* the two modes of being. The examples he gives involve the disruption of the plenitude of being in-itself by the nihilating activity of being in-itself, a disruption which introduces 'non-being' into the situation. Sartre gives the example of seeking his friend Pierre in a café, only to find that he is absent. This 'absence of Pierre' then becomes the central feature of the situation.[3] In general, the moment we refer to particular persons, things or situations, we disrupt the plenitude of being in-itself and introduce 'non-being' into it - I open my spectacle-case only to find the spectacles *not there*, but lying on a table across the room; or I reach into my pocket for my pen, which again is not there but lying on my desk; or again, I look for my shoes in the place I normally leave them, and find only one of them.

What is thus described is the reality of the world of our everyday concerns with people and things. This is no longer the world of an *undifferentiated* being in-itself, which by contrast is shown up as an abstraction, or at best a very special case of being in-itself, viz. the world as it would be for a consciousness that had no interest in it whatever, no project in relation to it, hence a world at the very limit of its 'worldhood'. If this is not too narrowly circular, the world as we encounter it is a world *for* consciousness, and consciousness is always a consciousness *for which* the world provides its possibilities for action, its significance, in short its very meaning. Sartre's account of consciousness as being for-itself is centred on 'freedom' understood as the act of 'nihilation' - a kind of distancing or standing back from any determinate phenomenon or element of being in-itself. It is further distinguished from being in-itself in that the latter embodies identity understood as 'sameness' - the chair is distinct from the table, the table from the floor, and each retains its identity, its sameness, in all contexts. Being for-itself, by contrast, is not 'identical' or 'the same' in this way, but involves instead a perpetual action of distancing, even of *self*-distancing.

'Freedom' is, for Sartre as it was for Hegel, the *central* characterisation of the project of consciousness, and the project necessarily takes a temporal form. Consciousness at an instant is an impossibility for

Sartre. It is easy to get the idea from reading Descartes, for example, that the subject of the 'cogito' could exist in an instant - that it could come into existence and go out of existence a moment later, and that between the two moments we would have consciousness in the full sense. Sartre not only denies this, but aims to show why it is impossible. In order genuinely to exist, consciousness as freedom has to have itself as its aim or its project, it has to *realise* itself as freedom, it has to create itself as a 'second nature', but of course without the essence-bound sense that the phrase 'second nature' might suggest in the context of, say, Aristotle's *Nicomachean Ethics*.[4] Hence it has to maintain itself in a way that is not tied down to just one particular instance of being in-itself - a consciousness tied down in this way would be like being locked into an experience like that of Roquentin in Sartre's *Nausea*, without any way out. Sartre's hero in the novel describes an experience of worldhood without meaning - a world dense with being, a world that contains no possibilities and invites no exercise of freedom.[5] Of course it is only a single experience - no consciousness could exist for which this was its permanent mode of being - but nonetheless it haunts the novel as a whole. Sartre's account of the temporality of consciousness provides us with both the necessary conditions for the possibility of freedom, and at the same time gives us reasons *why* a consciousness that fully corresponded to Roquentin's experience could not exist.

The project for consciousness is, then, that it has to create itself as freedom, and as a kind of 'second nature' that *embodies* that freedom. The concrete embodiment of freedom for consciousness, what it has always already accomplished, lies in its past, and what it is always free to do or to be lies in the future. The arena of the *enacting* of that freedom is the present. But this enacting of freedom never gives to consciousness the second nature it seeks where it most needs it, i.e. in the present, for as soon as it has acted, its actions cease to be the realisation of itself in the present world, as these actions slip into the past. As Sartre puts it, 'the past is a For-itself which has ceased to be a transcending presence to the In-itself. Now become in-itself, it has fallen *into the midst of the world*'.[6] By the very exercise of freedom as for-itself, I turn myself into being in-itself - and because the past is thus perpetually being pushed away from me, I am compelled perpetually to be free in the present, perpetually facing the future.

I am thus for ever seeking to attain some kind of solid being in the present, but the project is doomed to failure. No sooner have I acted, my actions cease to have the character of being for-itself - they peel off me into the world, into the past, and into the sphere of being in-itself. Although my freedom is perpetually being renewed in the present, there is a peculiar relentlessness about it - this perpetually-renewed freedom is, paradoxically, a perpetual bondage. I am for ever on the way to being, but this being that I seek for ever escapes me. The self's temporality for Sartre thus follows a kind of inner necessity, by which as soon as the self exercises its freedom in the present, as soon as the attainment of being seems imminent, that being slips away and is engulfed by the past. But the self is never able to *sustain* that past being which is its past self, never able to make it coextensive with its present self in any way that would accord with a conception of the self for which ground projects, character etc. are constitutive. As the past being of the for-itself becomes an object for it, and as its 'freedom' is thrust upon it again, so its present and its past being are distanced from each other - I never *am* my character or my ground projects, I merely 'have' them back there in the past as something I have not to be, in the nihilating act that is constitutive of being for-itself.

The advance on earlier theories lies, nonetheless, in the way the self is more radically implicated in its time. There is a way in which Sartre captures something of the temporality of the *agent* rather than merely that of the perceiver. For my past is not merely a picture-gallery of 'representations' but the real past I have been, the past I have lived through, the past as 'ecstasis' in the sense made familiar to us first through Heidegger, and then through Sartre, Merleau-Ponty and others.[7] The term comes from the Greek ἐκστατικόν, literally '(that which) stands out of itself'. I shall now attempt to paraphrase this notion in a way that engages with the main concerns of this essay.

The basic idea is that consciousness 'stands out of itself' in the sense that it is always already *engaged in* the world rather than merely *conscious of* it and contemplating it from ouside. In terms of time, Heidegger, hence also Sartre and Merleau-Ponty, speak of past, present and future as 'the three temporal *ekstasen*'. Consciousness engages with the present by 'standing out of itself' in such a way that it is inescapably and directly engaged with the world *itself* to which it is present, rather than indirectly and only cognitively, through 'representations'. The way consciousness engages with the past and the future is 'ecstatic' in the sense that the primordial significance of its past, or of its future, is again not a matter of representations and knowledge, but of a prereflective relation of *being* that engages with the past or the future itself. Consciousness, on this account, has a 'horizonal' character - necessarily it operates in an extended temporal framework involving past and future. Although I have not hitherto used the term 'horizon' in this connection, what it signifies is clearly contained in my earlier remarks about the temporality of ground projects and constitutive attachments, in section 2 of this chapter and in the second paragraph of this section.

A further point is that the basis for a notion of ecstatic temporality is in many respects already implicit in the analysis of 'the micro-structure of going on' that I undertook in chapter two section 4. One of the principal conclusions of that section was that when we speak of present events using the continuous present form, e.g. as in 'John is singing the blues', we should take seriously the thesis that we are reckoning with the just-past and the immediate future *themselves*, and not merely with representations of them. The subsequent discussion in sections 5 and 6 further supports the idea of temporal *ekstasen*. I have in mind here the notions of unsaturatedness, significance and emerging identity set in the context of section 5, where I argued for a reconciliation of the transitory and the extensive aspects of time, among other things. However, these earlier discussions do not exhaust the notion of ecstatic temporality. The notion only comes fully into its own in the context of a conception of the self.

We have so far looked at the temporal ecstases only through Sartre. The past for Sartre's being in-itself is all it has by way of determinate being - it has to be there, if anywhere, that character and ground projects are located. My past is all I have by way of my 'essence', who or what I am - but I 'am' it in only a very attenuated sense. The upsurge of the for-itself in the present can only cast my past self in the form of the for-itself that is no longer for-itself, but has become in-itself, 'fallen into the midst of the world' as Sartre puts it. What is of importance and value in Sartre's account of the temporality of consciousness, taking the past as our example, is the way he represents it as my past *being*, and not just a picture gallery of representations that I 'know through memory'. What is *wrong* with it, taking into account the discussion of Sartre in chapter three, is the way Sartre cannot allow it accommodate ground projects, character etc. without turning the self into a locus of self-deception or 'bad faith'. In order to avoid 'bad faith', I have to keep my 'past self', hence my character, ground projects and so on, perpetually at a distance from me as something I *have*, not something I *am*. I have to view my past self as if it were 'myself no longer'. Here again, as we saw in chapter two, we have privative language, privative metaphors, closing off the possibility of a different understanding. Although even for Sartre I have to be myself against the background of what I have been, the sense in which I *am* my past lies in my being for-others (être-pour-autrui). It is in the eyes of others in general that I have a particular character and a biography that enacts that character - from my own standpoint of consciousness and inescapable freedom, I *am not* but merely *have* a character with its ground projects. Hence on Sartre's 'fundamental ontology', I can never carry forward the ground projects which, for me, have their roots in the depths of my past. We need, therefore, to preserve the ecstatic account of temporality in a way that allows me to retain my past self as what I *am*, what I continually sustain, as what is open to my present self and capable of animating my present being. As will emerge presently, it is on Merleau-Ponty's account of consciousness and its temporality that this requirement is more readily accommodated.

4 Ecstatic Temporality and the Intra-Temporal 'I'

Sartre's account of consciousness leaves us with a picture of human life as the arena of choice, but where a certain *style* of choice is perpetually thrust upon the self. In the most general terms, the self is for ever engaged in choice *of* a world rather than choice *in* a world. Relatedly, I am never happy, but aware *of* happiness, and I am never angry, but aware *of* anger; in each case, I am aware of these as possible choices. Similarly, on Sartre's account Pasha Antipov ('Strelnikov') in *Doctor Zhivago* would have to be aware *of* his embracing of the Bolshevik revolutionary cause as a choice confronting him from moment to moment, as something always at a distance from him, rather than as something he has irrevocably made central to the ground project of his life, and on the basis of which further choices and decisions are made. What we need to say, following the extended discussion of ground projects, will and so on in chapter three, is that for Strelnikov choice is choice *within* a world, not choice *of* a world - it is choice within the world of his idealistic espousal of the Bolshevik revolutionary cause, and of his lofty and idealistic love for the Lara to whom he is married, but of whom he feels as yet unworthy.

In what follows, I address the issue of the conditions for the possibility of choice within, rather than of, a world. First, we have to do with ground projects properly understood, i.e. in accordance with the conception of the self that emerged in chapter three. Second, we have to see these ground projects as operating on an indefinitely large time-scale. This in turn entails that we have to take the ecstatic nature of temporality seriously. Third, we have to find a new way of understanding the identity of the self - one that entails neither a self persisting *in* time, nor a self standing *outside* time. This seems to require what I called, at the beginning of section 3, an 'immanent' or 'indwelling' principle, a principle according to which the self *structures* its time while at the same time *being* its time.

These last remarks may seem obscure. To clarify what I am trying to say, I shall begin with the observation that, for Sartre, the principle according to which the for-itself is never at home but perpetually in transit, perpetually compelled to escape its past being in order to seek a being in-itself it can never attain, is certainly a principle by which the self *structures* its time. But it is not a principle by which the self can *be* its time - or better, it can only be its time in the form of a freedom so absolute in its intent as to make it empty. For the self's project is freedom understood as freedom from every determinate being, a perpetually sustained affirmation of a freedom that has only itself as its object, hence a freedom without content. It is the perpetual 'ought-to-be' that never is, albeit *temporalised* in Sartre's characteristic way, and thus echoes Hegel's critique of Kant.[8] It is freedom in form only, precisely because it is freedom *from* a world, not freedom *in* a world: 'I am condemned to exist for ever beyond my essence, beyond the causes and motives of my act. I am condemned to be free. This means that no limits to my freedom can be found except freedom itself or, if you prefer, that we are not free to cease being free'.[9] In order to conceive *intra*-mundane freedom, we have to get beyond the privative account of the past, and bring the past self back into ecstatic identity with the present - hence not via 'representations' and 'memories', but in the primordial sense to which the notion of the past as an 'ekstase' speaks, and which Sartre, despite his intentions, covers up. My ground project, that which from the depths of my past constitutes the origin of what I will to be, is nothing other than my will in the widest sense, the sense familiar from chapter three. As such, it articulates itself through my time, forming the substance of my present self and propelling me forward into the future.

The indwelling principle we are seeking is, then, nothing other than my will in this wider sense, and we should recall at this stage something of what distinguishes it from the narrow conception of the will. It pervades the self as a whole, it is not directed to narrow, predescribed ends, it is as much to be characterised by acceptance and assent as by determination and 'choice'. It defines not simply what I typically *do*, but *how* I am and *what* I am, how I understand myself and my situation, and so on.

As I suggested at the end of the last section, it is Merleau-Ponty's rather than Sartre's account of the temporality of the self that can accommodate ground projects, character and so on. There are passages in *Phenomenology of Perception* that do not make sense unless we understand them as speaking to this view of the self, and such passages can be found as early as chapter 1 of the first section, 'The Body as Object and Mechanistic Physiology', as well as in the 'Temporality' chapter near the end of the book. In particular, it is possible to discern in these passages the wider notion of the will which, I claim, is the clue to the indwelling principle that propels the self out of its past and into its future:

> ...repression, to which psychoanalysis refers, consists in the subject's entering upon a certain course of action - a love affair, a career, a piece of work - in his encountering in this course some barrier, and, since he has the strength neither to surmount the obstacle nor to abandon the enterprise, he remains imprisoned in the attempt and uses up his strength indefinitely renewing it in spirit. Time in its passage does not carry away with it these impossible projects; it does not close up on the traumatic experience; the subject remains open to the same impossible future, if not in his explicit thoughts, at any rate in his actual being. One present among all presents thus acquires an exceptional value; it displaces the others and deprives them of their value as authentic presents. We continue to be the person who once entered on this adolescent affair, or the one who once lived in this parental universe. New perceptions, new emotions even, replace the old ones, but this process of renewal touches *only the content of our experience and not its structure*.[10]

The content may change, but the structure abides - I repeat what has now become an inseparable part of my ground project. Or it may be that the 'impossible project' to which Merleau-Ponty refers is itself a repetition of something else, something buried even more deeply in my past. He goes on to say how the 'structure' that abides remains hidden from view - in the cases which form the subject of Merleau-Ponty's example, the structure is never displayed before our gaze in memory, but remains hidden from view as the motivating source of the present; 'it is of its essence to survive only as a manner of being and with a certain degree of generality'.[11]

In the 'Temporality' chapter, Merleau-Ponty examines a passage from Proust's *Remembrance of Things Past* where we are given a description of Swann's love for Odette and the various transformations it undergoes. Merleau-Ponty objects to the way Proust speaks of Swann's love for Odette *causing* jealousy, and that jealousy *modifying* the love. It is a mistake to think of Swann's love *causing him to feel* jealousy, because it *is* and *was* jealousy from the start: 'Swann's feeling of pleasure in looking at Odette bore its degeneration within itself, since it was the pleaure of being the only one to do so'.[12] He comments further that Swann's jealous love should be related to the rest of his behaviour, from which it might be learned that this love manifests something of the whole of Swann's personality. This is clearly the language of ground projects, and of will in the wider sense. But nothing of this, he goes on to add, is comprehensible if we think of the 'I' as a 'thinking' or 'constituting' subject:

> Here is where temporality throws light on subjectivity. We shall never manage to understand how a thinking or constituting subject is able to posit or become aware of itself in time. If the *I* is indeed the transcendental Ego of Kant, we shall never understand how it can in any instance merge with its wake in the inner sense, or how the empirical still remains a self. If, however, the subject is identified with temporality, then self-positing ceases to be a contradiction, because it expresses the essence of living time'.[13]

The fact that he speaks of the intra-temporal being of the 'I' in this way clearly marks his fundamental difference with Sartre. It is interesting to note further that, in contrast wirh Sartre, Merleau-Ponty speaks here of the '*I*' and not the '*me*'. For precisely what Sartre's for-itself does is to constitute the '*me*' and then let it fall into the midst of being in-itself, just as Kant's transcendental self constitutes the phenomenal self in the inner sense.

The question remains how this intra-temporal 'I' is possible. We have seen that it must be the will in the widest sense. For as we saw in chapter three, the will understood in the widest sense, the sense I have been advocating in this essay, is not something that issues in a set of discrete 'acts', but a total condition of the self. It goes without saying that it makes no sense - it is a 'category mistake', perhaps - to speak of 'will' in this sense as sometimes present, sometimes absent. What we must *not* say is that the 'I' *produces* this intra-temporal will - for then the temptation will always be there to re-invent the transcendental ego we have banished. Only by acknowledging that the 'I' simply *is* the intra-temporal will can we have a conception of temporality that is adequate to the redrawn self. It remains for us to draw out what follows from this overall conclusion.

Clearly the very idea of an intra-temporal 'I' understood in this way would be, for Sartre, a colossal exercise in bad faith. It would be to deny the freedom of consciousness as he understands it. But such an understanding of freedom is, I have argued, illusory; it is a freedom that manifests only as choice *of* a world from outside that world, not choice within the world. Of course if the 'I' *is* intra-temporal, it is vulnerable - it is not simply *aware* of anxiety or depression, it *is* anxious or depressed. One commentary on Sartre points to the way his account of consciousness falsifies experience, in that it always portrays consciousness as distanced, as lacking something, and as being perpetually prone to bad faith. The authors point to the way the depressed person, for example, may experience no distance between himself and his depression - he is, in his very being, depressed; he does not merely observe depression. But the authors also point to more positive experiences, with the example of a virtuoso cellist for whom there is no distance between consciousness and the music.[14]

The possibility of these experiences, negative or positive, can be understood only on the more substantial conception of the self I have advocated in this and the last chapter. In particular, I draw attention to the discussion of the will in the last chapter, and especially to Hegel's account of the will.[15] For Hegel's account speaks to an understanding of the self as *engaged* in its projects rather than distanced from them - an understanding of the self for which its projects are therefore constitutive. If the self is engaged in its projects in this way, then it is not distanced from its past - least pof all in the manner we find in Sartre's account of consciousness. The past as 'ecstatic' is conceivable only if it is possible for the self to be engaged in its projects in this manner. This has, inevitably, its negative and its positive sides. Merleau-Ponty points to the negative experience whereby I may be vulnerable to my past in a way that threatens to draw energy away from me:: 'we never completely extricate ourselves from it, time never completely closes over it and it remains like a wound through which our strength ebbs away'.[16] An example of this would be a decision I made long ago that I have never stopped regretting, and which draws my strength away whenever I chance to think about it. But equally one can return to a past event that is a source of strength, such as a particularly fortunate and fulfilled part of one's life. To be capable of joy, we must be capable of suffering.

To bring this discussion to a close, I make one final point. At the end of chapter three, it seemed both that the case for 'redrawing the self' was overwhelming, and that we could not have the benefit of a picture, like the S→O schema, that would encapsulate the outcome. The self, it was concluded, has to be redrawn in words. To seek out the temporality of the self thus redrawn has been the principal aim of this chapter. The idea of an intra-temporal 'I' receives its justification in the context of chapter three from the necessity for steering between the Scylla of the transcendental ego on the one side, and the Charybdis of what Sandel calls 'a radically situated subject', i.e. a subject that is *nothing at all* beyond the series of its experiences, on the other.[17]

Clearly the intra-temporal 'I' thus understood sits uncomfortably next to a metaphysics of time that is concerned primarily with persistence through time, change and the permanent-through-change, and so on. It is for this reason that its temporality has to be 'ecstatic' - it has not simply to *have* a past, a present and a future, it has to *be* its past, its present and its future. To show how this is possible has been the main aim of this chapter.

What finally needs to be said here is that a self that has to be its past - *has to be* it, without Sartre's addition 'in the mode of not being it' - does, in the end, lend weight to the philosophical preoccupation with getting beyond the 'thin' conception of the self familiar to us through the epistemological tradition that runs from Descartes to Kant and beyond. The intra-temporal 'I' is therefore inescapable.

5 Concluding Remarks

In the course of this essay, we have viewed time as an object *for* consciousness - the time that is witnessed, time *for* the self - and time as constitutive *of* consciousness, the time *of* the self. How do these different temporal discourses relate to each other? I close with some thoughts on this question.

In chapter two, we found that time as an object for consciousness is best viewed through the notions of unsaturatedness, significance, emerging identity and narrative. We also concluded that there is no restriction *a priori* on what is to count as a present event, series of events or process. But there is more to ecstatic temporality than these earlier discussions might suggest. Time as presented to consciousness is grounded in the intra-temporal 'I' and its ecstatic temporality, and not vice versa.

I *am* my past, my present and my future - I do not merely observe them. The world as object of consciousness, as when I hear a symphony or an opera, does not in any way share in this ecstatic character, because I am not the symphony or the opera no matter how deep my enjoyment of them. But what the two temporal discourses share is the extended future, present and past horizon. Both are 'horizonal', though one of them is also 'ecstatic' in the sense explained earlier in this chapter. The two discourses meet in the fact that we are not merely perceivers of a world, but agents engaged in it. If the overall argument of this essay is right, the fact that we are engaged in the world, and engaged in it 'ecstatically', is *the* basic fact of our situation. The reflective and contemplative aspects of our lives, as when we listen to music or watch a performance, are possibilities that are developed out of that prior, basic fact - possibilities that are therefore grounded in the sphere of our prereflective being, the sphere of the intra-temporal 'I' with its ecstatic temporality.

Notes and References

[1]   See chapter two, p. 45 and pp. 49-52.

[2]   J-P. Sartre, *Being and Nothingness* (trans. Hazel Barnes), London, 1966, Introduction p. lxvii.

[3]   Ibid., pp. 9-10.

[4]   Aristotle does not use the phrase 'second nature' as such - but his discussion of the ethical virtues is strongly suggestive of it, in that he says that the virtues are in us neither by nature nor against nature, but rather nature gives us the capacity to receive them. Aristotle, *Nicomachean Ethics*, trans. H. Rackham, Cambridge Mass. (USA) and London (UK) 1982, II i, 3-4.

[5]   I refer here to the well-known passage where Roquentin describes such an experience in a municipal park, staring at a large tree-root. See J-P. Sartre, *Nausea*, trans. R. Baldwick, Harmondsworth 1965, pp. 182-193.

[6]   Sartre, *Being and Nothingness*, p. 146 (Sartre's italics).

[7]   See M. Heidegger, *Being and Time*, trans. J. Macquarrie and E. Robinson, Oxford 1967, p. 377 (H 329); also Sartre, *Being and Nothingness*, part 2 chapter 2 passim; and also M. Merleau-Ponty, *Phenomenology of Perception*, London 1965, part three chapter 2 passim..

[8]   See Hegel, *Philosophy of Right*, Second Part: Morality, sub-section 3, passim..

[9]   Sartre, *Being and Nothingness*, p. 439.

[10]   Merleau-Ponty, *Phenomenology of Perception*, p. 83. (My italics).

[11]   Idid., p. 83.

[12]   Ibid., p. 425.

[13]   Ibid., p. 425.

[14]   See M. Hammond, J. Howarth and R. Keat, *Understanding Phenomenology*, Oxford, 1991, pp. 125-6.

[15]   See pp. 70-79 above.

[16]   Merleau-Ponty, *Phenomenology of Perception*, p. 85.

[17]   See M. Sandel, *Liberalism and the Limits of Justice*, Cambridge, 1982, pp. 20-21.

Bibliography

Aristotle, *Politics*, trans. H. Rackham, Cambridge Massachusetts and London, 1977.

Aristotle, *Nicomachean Ethics*, trans. H. Rackham, Cambridge Massachusetts and London, 1982.

St. Augustine, *Sancti Aurelii Augustini (Hipponensis Episcopi) Opera Omnia*, Paris, 1836.

St. Augustine, *Confessions*, trans. R.S. Pine-Coffin, Harmondsworth, 1961.

Robert Bolt, *A Man For All Seasons*, London, 1960.

F.H. Bradley, *Ethical Studies*, Oxford, 1927.

C.D. Broad, *An Examination of McTaggart's Philosophy*, Cambridge, 1938.

D. Davidson, *Essays on Actions and Events*, Oxford, 1980.

N.J.H. Dent, 'Plato and Social Justice', in Andros Loizou and Harry Lesser (eds.), *Polis and Politics* (cited below).

R.M. Gale, *The Language of Time*, London, 1968.

R.M. Gale (ed.), *The Philosophy of Time*, London and Melbourne, 1968.

R.W. Hall, 'Platonic Justice and the *Republic*', *Polis*, Volume 6 No. 2, 1987.

R.W. Hall, 'Plato and Totalitarianism', *Polis*, Volume 7 No. 2, 1988.

M. Hammond, J. Howarth and R. Keat, *Understanding Phenomenology*, Oxford, 1991.

S.N. Hampshire, *Spinoza*, Harmondsworth, 1951.

G.W.F. Hegel, *Philosophy of Right*, trans. T.M. Knox, Oxford, 1967.

M. Heidegger, *Being and Time*, trans. J. Macquarrie and E. Robinson, Oxford, 1967.

D. Hume, *A Treatise of Human Nature*, L.A. Selby-Bigge (ed.), Oxford, 1960.

E. Husserl, *The Phenomenology of Internal Time-Consciousness*, trans. James S. Churchill, The Hague, 1964.

I. Kant, *Critique of Pure Reason*, trans. Norman Kemp Smith, London, 1961.

R. Le Poidevin, *Change, Cause and Contradiction: A Defence of the Tenseless Theory of Time*, Basingstoke and London, 1991.

A.H. Lesser, 'Society and Individual Psychology in Plato's *Republic*', in Andros Loizou and Harry Lesser (eds.), *Polis and Politics* (cited below).

A. Loizou, *The Reality of Time*, Aldershot and Brookfield (USA), 1986.

A. Loizou, '*Eikasia* and Tyranny in Plato's *Republic*', in Andros Loizou and Harry Lesser (eds.), *Polis and Politics* (cited below).

A. Loizou, 'Scepticism, Embodied Existence and Time', in *Skepsis VIII*, Athens and Olympia, 1997.

A. Loizou, 'The Threefold Psyche and the Dramatisation of Justice in Plato's *Republic*', in *Polis*, Volume 16, 1999.

A. Loizou and A.H. Lesser (eds.), *Polis and Politics: Essays in Greek Moral and Political Philosophy*, Aldershot and Brookfield (USA), 1990.

D. McLellan (ed.), *Karl Marx: Early Texts*, Oxford, 1979.

J.M.E. McTaggart, 'The Unreality of Time', in J.M.E. McTaggart, *Philosophical Studies*, London, 1934.

J.M.E. McTaggart, *The Nature of Existence*, Cambridge, 1968.

D.H. Mellor, *Real Time II*, London and New York, 1998.

M. Merleau-Ponty, *Phenomenology of Perception*, trans. Colin Smith, London, 1965.

G.E. Moore, *The Commonplace Book of G.E. Moore*, Casimir Lewy (ed.), London, 1962.

Wilfred Owen, *The Collected Poems of Wilfred Owen*, C. Day Lewis (ed.), London, 1963.

B. Pasternak, *Doctor Zhivago*, trans. Max Hayward and Manya Harari, London, 1958.

D.F. Pears, 'Time, Truth and Inference', in A. Flew (ed.), *Essays in Conceptual Analysis*, London, 1966.

Plato, *Republic*, trans. Paul Shorey, Cambridge Massachusetts and London, 1978.

M. Proust, *Remembrance of Things Past*, trans. C.K. Scott Moncrieff and Terence Kilmartin, Harmondworth, 1989.

H. Putnam, *Meaning and the Moral Sciences*, London, 1978.

W.V.O. Quine, *Word and Object*, Cambridge Massachusetts, 1960.

G. Ryle, *The Concept of Mind*, London, 1949.

M. Sandel, *Liberalism and the Limits of Justice*, Cambridge, 1982.

J-P. Sartre, *Existentialism and Humanism*, trans. P. Mairet, London, 1948.

J-P. Sartre, *Being and Nothingness*, trans. Hazel Barnes, London, 1966.

J-P. Sartre, *Nausea*, trans. Robert Baldick, Harmondsworth, 1972.

J-P. Sartre, *The Transcendence of the Ego*, Trans. Forrest Williams and Robert Kirkpatrick, New York, 1975.

S. Shoemaker, *Self-Knowledge and Self-Identity*, Ithaca, New York, 1963.

B. Spinoza, *Ethics*, trans. Andrew Boyle, London and New York, 1959.

P.F. Strawson, *Individuals: An Essay in Descriptive Metaphysics*, London, 1959.

P.F. Strawson, *The Bounds of Sense: An Essay on Kant's Critique of Pure Reason*, London, 1966.

C.C.W. Taylor, 'Plato's Totalitarianism', in *Polis*, Volume 5 No. 2, 1986.

L. Tolstoy, *Anna Karenin*, trans. Rosemary Edmonds, Harmondsworth, 1972.

B.A.O. Williams, 'Persons, Character and Morality', in A.O. Rorty (ed.), *The Idenities of Persons*, Berkeley, Los Angeles, London, 1976.

L. Wittgenstein, *Philosophical Investigations*, Oxford, 1963.

D. Wood, *The Deconstruction of Time*, Atlantic Highlands, NJ, 1989.

Index